S0-BNA-775

DISCARDED

U•X•L Encyclopedia
of Biomes

Second Edition

U·X·L Encyclopedia of Biomes

Second Edition VOLUME 2

Marlene Weigel

U·X·L

A part of Gale, Cengage Learning

GALE
CENGAGE Learning

Detroit • New York • San Francisco • New Haven, Conn • Waterville, Maine • London

U·X·L Encyclopedia of Biomes
Marlene Weigel

Project Editor: Madeline Harris

Editorial: Kathleen Edgar, Debra Kirby, Kristine Krapp, Kimberley McGrath, and Lemma Shomali

Composition: Evi Abou-El-Seoud

Imaging: Lezlie Light

Manufacturing: Wendy Blurton

Product Design: Jennifer Wahi

Product Management: Julia Furtaw

Rights Acquisition and Management: Dean Dauphinais and Robyn Young

© 2010 Gale, Cengage Learning

ALL RIGHTS RESERVED. No part of this work covered by the copyright herein may be reproduced, transmitted, stored, or used in any form or by any means graphic, electronic, or mechanical, including but not limited to photocopying, recording, scanning, digitizing, taping, Web distribution, information networks, or information storage and retrieval systems, except as permitted under Section 107 or 108 of the 1976 United States Copyright Act, without the prior written permission of the publisher.

For product information and technology assistance, contact us at **Gale Customer Support, 1-800-877-4253.**
For permission to use material from this text or product, submit all requests online at **www.cengage.com/permissions.**
Further permissions questions can be emailed to **permissionrequest@cengage.com**

Cover photographs: Image copyright 2009: Debra Hughes (tree icon), Alexander Kolomietz (beach scene), Sergey Popov (baracudas), Snowleopard1 (frog), Chee-Onn Leong (tall pines), Sigen Photography (desert), and Oksana Perkins (pond and mountains), all used under licence from Shutterstock.com. Polar bear and fall leaves images © 2009/Getty Images.

While every effort has been made to ensure the reliability of the information presented in this publication, Gale, a part of Cengage Learning, does not guarantee the accuracy of the data contained herein. Gale accepts no payment for listing; and inclusion in the publication of any organization, agency, institution, publication, service, or individual does not imply endorsement of the editors or publisher. Errors brought to the attention of the publisher and verified to the satisfaction of the publisher will be corrected in future editions.

Library of Congress Cataloging-in-Publication Data

U-X-L encyclopedia of biomes / [editor] Marlene Weigel. -- 2nd ed.
    p. cm. --
    Includes bibliographical references and index.
    ISBN 978-1-4144-5516-7 (set) -- ISBN 978-1-4144-5517-4 (vol. 1) -- ISBN978-1-4144-5518-1 (vol. 2) -- ISBN 978-1-4144-5519-8 (vol. 3) -- ISBN 978-1-4144-4426-0 (e-book)
    1. Biotic communities--Juvenile literature. I. Weigel, Marlene. II. Title: Encyclopedia of biomes.
    QH541.14.U18 2009
    577.8'2--dc22   2008014502

Gale
27500 Drake Rd.
Farmington Hills, MI, 48331-3535

978-1-4144-5516-7 (set)   1-4144-5516-X (set)
978-1-4144-5517-4 (vol. 1)   1-4144-5517-8 (vol. 1)
978-1-4144-5518-1 (vol. 2)   1-4144-5518-6 (vol. 2)
978-1-4144-5519-8 (vol. 3)   1-4144-5519-4 (vol. 3)

This title is also available as an e-book.
ISBN-13: 978-1-4144-4426-0 ISBN-10: 1-4144-4426-5
Contact your Gale, a part of Cengage Learning, sales representative for ordering information.

Printed in China by China Translation & Printing Services Limited
1 2 3 4 5 6 7 13 12 11 10 09

Table of Contents

Reader's Guide

This second edition of U•X•L Encyclopedia of Biomes offers readers comprehensive, easy-to-use, and current information on twelve of Earth's major biomes and their many component ecosystems. Arranged alphabetically across three volumes, each biome chapter includes: an overview; a description of how the biomes are formed; their climate; elevation; growing season; plants, animals, and endangered species; food webs; human culture; and economy. The information presented may be used in a variety of subject areas, such as biology, geography, anthropology, and current events. Each chapter includes a "spotlight" feature focusing on specific geographical areas related to the biome being discussed and concludes with a section composed of books, periodicals, Internet addresses, and environmental organizations for readers to conduct more extensive research.

Additional Features

Each volume of *U•X•L Encyclopedia of Biomes* includes color maps and at least 60 photos and illustrations pertaining to each biome, while sidebar boxes highlight fascinating facts and related information. All three volumes include a glossary, a bibliography, and a subject index covering all the subjects discussed in *U•X•L Encyclopedia of Biomes*.

Note

There are many different ways to describe certain aspects of a particular biome, and it would be impossible to include all of the classifications in *U•X•L Encyclopedia of Biomes*. However, in cases where more than one classification seemed useful, more than one was given. Please note that the classifications represented here may not be those preferred by all specialists in a particular field.

Every effort was made in this set to include the most accurate information pertaining to areas and other measurements. Great variations exist in the available data, however. Sometimes differences can be accounted for in terms of what was measured: the reported area of a lake, for example, may vary depending upon the point at which measuring began. Other differences may result from natural changes that took place between the time of one measurement and another. Further, other data may be questionable because reliable information has been difficult to obtain. This is particularly true for remote areas in developing countries, where funds for scientific research are lacking and non-native scientists may not be welcomed.

The U•X•L editors would like to thank author Marlene Weigel for her work on the original edition of this title. We also thank contributing writer Rita Travis for her work on the Coniferous Forest, Grassland, Tundra, and Wetland chapters. All the entries were updated by Lubnah Shomali for this new edition, and we relied heavily on our expert, Dr. Dan Skean, of Albion College to provide academic insight and additional feedback.

Comments and Suggestions

We welcome your comments on this work as well as your suggestions for topics to be featured in future editions of *U•X•L Encyclopedia of Biomes* Please write: Editors, *U•X•L Encyclopedia of Biomes,* U•X•L, 27500 Drake Rd., Farmington Hills, MI 48331-3535; call toll-free: 1-800-877-4253; fax: 248-699-8097; or send e-mail via www.gale.cengage.com.

Biomes of the World Map

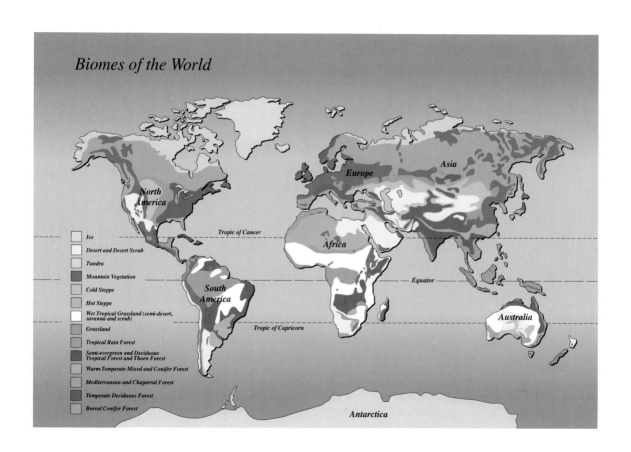

Biomes of the World

Ice
Desert and Desert Scrub
Tundra
Mountain Vegetation
Cold Steppe
Hot Steppe
Wet Tropical Grassland (semi-desert, savanna and scrub)
Grassland
Tropical Rain Forest
Semi-evergreen and Deciduous Tropical Forest and Thorn Forest
Warm Temperate Mixed and Conifer Forest
Mediterranean and Chaparral Forest
Temperate Deciduous Forest
Boreal Conifer Forest

North America

South America

Europe

Asia

Africa

Australia

Antarctica

Tropic of Cancer

Equator

Tropic of Capricorn

Words to Know

Abyssal plain: The flat midportion of the ocean floor that begins beyond the continental rise.

Acid rain: A mixture of water vapor and polluting compounds in the atmosphere that falls to Earth as rain or snow.

Active margin: A continental margin constantly being changed by earthquake and volcanic action.

Aerial roots: Plant roots that dangle in midair and absorb nutrients from their surroundings rather than from the soil.

Algae: Plantlike organisms that usually live in watery environments and depend upon photosynthesis for food.

Algal blooms: Sudden increases in the growth of algae on the ocean's surface.

Alluvial fan: A fan-shaped area created when a river or stream flows downhill, depositing sediment into a broader base that spreads outward.

Amphibians: Animals that spend part, if not most, of their lives in water.

Amphibious: Able to live on land or in water.

Angiosperms: Trees that bear flowers and produce their seeds inside a fruit; deciduous and rain forest trees are usually angiosperms.

Annuals: Plants that live for only one year or one growing season.

Aquatic: Having to do with water.

Aquifier: Rock beneath Earth's surface in which groundwater is stored.

Arachnids: Class of animals that includes spiders and scorpions.

Arctic tundra: Tundra located in the far north, close to or above the Arctic Circle.

Arid: Dry.

Arroyo: The dry bed of a stream that flows only after rain; also called a wash or a _wadi._

Artifacts: Objects made by humans, including tools, weapons, jars, and clothing.

Artificial grassland: A grassland created by humans.

Artificial wetland: A wetland created by humans.

Atlantic blanket bogs: Blanket bogs in Ireland that are less than 656 feet (200 meters) above sea level.

Atolls: Ring-shaped reefs formed around a lagoon by tiny animals called corals.

Bactrian camel: The two-humped camel native to central Asia.

Bar: An underwater ridge of sand or gravel formed by tides or currents that extends across the mouth of a bay.

Barchan dunes: Sand dunes formed into crescent shapes with pointed ends created by wind blowing in the direction of their points.

Barrier island: An offshore island running parallel to a coastline that helps shelter the coast from the force of ocean waves.

Barrier reef: A type of reef that lines the edge of a continental shelf and separates it from deep ocean water. A barrier reef may enclose a lagoon and even small islands.

Bathypelagic zone: An oceanic zone based on depth that ranges from 3,300 to 13,000 feet (1,000 to 4,000 meters).

Bathyscaphe: A small, manned, submersible vehicle that accommodates several people and is able to withstand the extreme pressures of the deep ocean.

Bay: An area of the ocean partly enclosed by land; its opening into the ocean is called a mouth.

Beach: An almost level stretch of land along a shoreline.

Bed: The bottom of a river or stream channel.

Benthic: Term used to describe plants or animals that live attached to the seafloor.

Biodiverse: Term used to describe an environment that supports a wide variety of plants and animals.

Bio-indicators: Plants or animals whose health is used to indicate the general health of their environment.

Biological productivity: The growth rate of life forms in a certain period of time.

Biome: A distinct, natural community chiefly distinguished by its plant life and climate.

Blanket bogs: Shallow bogs that spread out like a blanket; they form in areas with relatively high levels of annual rainfall.

Bog: A type of wetland that has wet, spongy, acidic soil called peat.

Boreal forest: A type of coniferous forest found in areas bordering the Arctic tundra. Also called taiga.

Boundary layer: A thin layer of water along the floor of a river channel where friction has stopped the flow completely.

Brackish water: A mixture of freshwater and saltwater.

Braided stream: A stream consisting of a network of interconnecting channels broken by islands or ridges of sediment, primarily mud, sand, or gravel.

Branching network: A network of streams and smaller rivers that feeds a large river.

Breaker: A wave that collapses on a shoreline because the water at the bottom is slowed by friction as it travels along the ocean floor and the top outruns it.

Browsers: Herbivorous animals that eat from trees and shrubs.

Buoyancy: Ability to float.

Buran: Strong, northeasterly wind that blows over the Russian steppes.

Buttresses: Winglike thickenings of the lower trunk that give tall trees extra support.

Canopy: A roof over the forest created by the foliage of the tallest trees.

Canyon: A long, narrow valley between high cliffs that has been formed by the eroding force of a river.

Carbon cycle: Natural cycle in which trees remove excess carbon dioxide from the air and use it during photosynthesis. Carbon is then returned to the soil when trees die and decay.

Carnivore: A meat-eating plant or animal.

Carrion: Decaying flesh of dead animals.

Cay: An island formed from a coral reef.

Channel: The path along which a river or stream flows.

Chemosynthesis: A chemical process by which deep-sea bacteria use organic compounds to obtain food and oxygen.

Chernozim: A type of temperate grassland soil; also called black earth.

Chinook: A warm, dry wind that blows over the Rocky Mountains in North America.

Chitin: A hard chemical substance that forms the outer shell of certain invertebrates.

Chlorophyll: The green pigment in leaves used by plants to turn energy from the sun into food.

Clear-cutting: The cutting down of every tree in a selected area.

Climax forest: A forest in which only one species of tree grows because it has taken over and only that species can survive there.

Climbers: Plants that have roots in the ground but use hooklike tendrils to climb on the trunks and limbs of trees in order to reach the canopy, where there is light.

Cloud forest: A type of rain forest that occurs at elevations over 10,500 feet (3,200 meters) and that is covered by clouds most of the time.

Commensalism: Relationship between organisms in which one reaps a benefit from the other without either harming or helping the other.

Commercial fishing: Fishing done to earn money.

Conifer: A tree that produces seeds inside cones.

Coniferous trees: Trees, such as pines, spruces, and firs, that produce seeds within a cone.

Consumers: Animals in the food web that eat either plants or other animals.

Continental shelf: A flat extension of a continent that tapers gently into the sea.

Continental slope: An extension of a continent beyond the continental shelf that dips steeply into the sea.

Convergent evolution: When distantly related animals in different parts of the world evolve similar characteristics.

Coral reef: A wall formed by the skeletons of tiny animals called corals.

Coriolis effect: An effect on wind and current direction caused by Earth's rotation.

Crustaceans: Invertebrate animals that have hard outer shells.

Current: The steady flow of water in a certain direction.

Dambo: Small marsh found in Africa.

Dark zone: The deepest part of the ocean, where no light reaches.

Deciduous: Term used to describe trees, such as oaks and elms, that lose their leaves during cold or very dry seasons.

Decompose: The breaking down of dead plants and animals in order to release nutrients back into the environment.

Decomposers: Organisms that feed on dead organic materials, releasing nutrients into the environment.

Deforestation: The cutting down of all the trees in a forest.

Dehydration: Excessive loss of water from the body.

Delta: Muddy sediments that have formed a triangular shape over the continental shelf near the mouth of a river.

Deposition: The carrying of sediments by a river from one place to another and depositing them.

Desalination: Removing the salt from seawater.

Desert: A very dry area receiving no more than 10 inches (25 centimeters) of rain during a year and supporting little plant or animal life.

Desertification: The changing of fertile lands into deserts through destruction of vegetation (plant life) or depletion of soil nutrients. Topsoil and groundwater are eventually lost as well.

Desert varnish: A dark sheen on rocks and sand believed to be caused by the chemical reaction between overnight dew and minerals in the soil.

Diatom: A type of phytoplankton with a geometric shape and a hard, glasslike shell.

Dinoflagellate: A type of phytoplankton having two whiplike attachments that whirl in the water.

Discharge: The amount of water that flows out of a river or stream into another river, a lake, or the ocean.

Doldrums: Very light winds near the equator that create little or no movement in the ocean.

Downstream: The direction toward which a river or stream is flowing.

Drainage basin: All the land area that supplies water to a river or stream.

Dromedary: The one-humped, or Arabian, camel.

Drought: A long, extremely dry period.

Dune: A hill or ridge of sand created by the wind.

Duricrusts: Hard, rocklike crusts on ridges that are formed by a chemical reaction caused by the combination of dew and minerals such as limestone.

Ecosystem: A network of organisms that have adapted to a particular environment.

Elfin forest: The upper cloud forest at about 10,000 feet (3,000 meters) which has trees that tend to be smaller, and twisted.

Emergents: The trees that stand taller than surrounding trees.

Epiphytes: Plants that grow on other plants with their roots exposed to the air. Sometimes called "air" plants.

Etermy: A current that moves against the regular current, usually in a circular motion.

Elevation: The height of an object in relation to sea level.

Emergent plants: Plants that are rooted at the bottom of a body of water that have top portions that appear to be above the water's surface.

Emergents: The very tallest trees in the rain forest, which tower above the canopy.

Engineered wood: Manufactured wood products composed of particles of several types of wood mixed with strong glues and preservatives.

Epilimnion: The layer of warm or cold water closest to the surface of a large lake.

Epipelagic zone: An oceanic zone based on depth that reaches down to 650 feet (200 meters).

Epiphytes: Plants that grow on other plants or hang on them for physical support.

Ergs: Arabian word for vast seas of sand dunes, especially those found in the Sahara.

Erosion: Wearing away of the land.

Estivation: An inactive period experienced by some animals during very hot months.

Estuary: The place where a river traveling through lowlands meets the ocean in a semi-enclosed area.

Euphotic zone: The zone in a lake where sunlight can reach.

Eutrophication: Loss of oxygen in a lake or pond because increased plant growth has blocked sunlight.

Fast ice: Ice formed on the surface of the ocean between pack ice and land.

Faults: Breaks in Earth's crust caused by earthquake action.

Fell-fields: Bare rock-covered ground in the alpine tundra.

Fen: A bog that lies at or below sea level and is fed by mineral-rich ground-water.

First-generation stream: The type of stream on which a branching network is based; a stream with few tributaries. Two first-generation streams join to form a second-generation stream and so on.

Fish farms: Farms in which fish are raised for commercial use; also called hatcheries.

Fjords: Long, narrow, deep arms of the ocean that project inland.

Flash flood: A flood caused when a sudden rainstorm fills a dry riverbed to overflowing.

Floating aquatic plant: A plant that floats either partly or completely on top of the water.

Flood: An overflow caused when more water enters a river or stream than its channel can hold at one time.

Floodplain: Low-lying, flat land easily flooded because it is located next to streams and rivers.

Food chain: The transfer of energy from organism to organism when one organism eats another.

Food web: All of the possible feeding relationships that exist in a biome.

Forbs: A category of flowering, broadleaved plants other than grasses that lack woody stems.

Forest: A large number of trees covering not less than 25 percent of the area where the tops of the trees interlock, forming a canopy at maturity.

Fossil fuels: Fuels made from oil and gas that formed over time from sediments made of dead plants and animals.

Fossils: Remains of ancient plants or animals that have turned to stone.

Freshwater lake: A lake that contains relatively pure water and relatively little salt or soda.

Freshwater marsh: A wetland fed by freshwater and characterized by poorly drained soil and plant life dominated by nonwoody plants.

Freshwater swamp: A wetland fed by freshwater and characterized by poorly drained soil and plant life dominated by trees.

Friction: The resistance to motion when one object rubs against another.

Fringing reef: A type of coral reef that develops close to the land; no lagoon separates it from the shore.

Frond: A leaflike organ found on all species of kelp plants.

Fungi: Plantlike organisms that cannot make their own food by means of photosynthesis; instead they grow on decaying organic matter or live as parasites on a host.

Geyser: A spring heated by volcanic action. Some geysers produce enough steam to cause periodic eruptions of water.

Glacial moraine: A pile of rocks and sediments created as a glacier moves across an area.

Global warming: Warming of Earth's climate that may be speeded up by air pollution.

Gorge: A deep, narrow pass between mountains.

Grassland: A biome in which the dominant vegetation is grasses rather than trees or tall shrubs.

Grazers: Herbivorous animals that eat low-growing plants such as grass.

Ground birds: Birds that hunt food and make nests on the ground or close to it.

Groundwater: Freshwater stored in rock layers beneath the ground.

Gulf: A large area of the ocean partly enclosed by land; its opening is called a strait.

Gymnosperms: Trees that produce seeds that are often collected together into cones; most conifers are gymnosperms.

Gyre: A circular or spiral motion.

Hadal zone: An oceanic zone based on depth that reaches from 20,000 to 35,630 feet (6,000 to 10,860 meters).

Hardwoods: Woods usually produced by deciduous trees, such as oaks and elms.

Hatcheries: Farms in which fish are raised for commercial use; also called fish farms.

Headland: An arm of land made from hard rock that juts out into the ocean after softer rock has been eroded away by the force of tides and waves.

Headwaters: The source of a river or stream.

Herbicides: Poisons used to control weeds or any other unwanted plants.

Herbivore: An animal that eats only plant matter.

Herders: People who raise herds of animals for food and other needs; they may also raise some crops but are usually not dependent upon them.

Hermaphroditic: Term used to describe an animal or plant in which reproductive organs of both sexes are present in one individual.

Hibernation: An inactive period experienced by some animals during very cold months.

High tide: A rising of the surface level of the ocean caused by Earth's rotation and the gravitational pull of the sun and moon.

Holdfast: A rootlike structure by which kelp plants anchor themselves to rocks or the seafloor.

Hummocks: 1. Rounded hills or ridges, often heavily wooded, that are higher than the surrounding area; 2. Irregularly shaped ridges formed when large blocks of ice hit each other and one slides on top of the other; also called hammocks.

Humus: The nutrient-rich, spongy matter produced when the remains of plants and animals are broken down to form soil.

Hunter-gatherers: People who live by hunting animals and gathering nuts, berries, and fruits; normally, they do not raise crops or animals.

Hurricane: A violent tropical storm that begins over the ocean.

Hydric soil: Soil that contains a lot of water but little oxygen.

Hydrologic cycle: The manner in which molecules of water evaporate, condense as clouds, and return to Earth as precipitation.

Hydrophytes: Plants that are adapted to grow in water or very wet soil.

Hypolimnion: The layer of warm or cold water closest to the bottom of a large lake.

Hypothermia: A lowering of the body temperature that can result in death.

Insecticides: Poisons that kill insects.

Insectivores: Plants and animals that feed on insects.

Intermittent stream: A stream that flows only during certain seasons.

Interrupted stream: A stream that flows aboveground in some places and belowground in others.

Intertidal zone: The seashore zone covered with water during high tide and dry during low tide; also called the middle, or the littoral, zone.

Invertebrates: Animals without a backbone.

Kelp: A type of brown algae that usually grows on rocks in temperate water.

Kettle: A large pit created by a glacier that fills with water and becomes a pond or lake.

Kopjes: Small hills made out of rocks that are found on African grasslands.

Labrador Current: An icy Arctic current that mixes with warmer waters off the coast of northeastern Canada.

Lagoon: A large pool of seawater cut off from the ocean by a bar or other landmass.

Lake: A usually large body of inland water that is deep enough to have two distinct layers based on temperature.

Latitude: A measurement on a map or globe of a location north or south of the equator. The measurements are made in degrees, with the equator, or dissecting line, being zero.

Layering: Tree reproduction that occurs when a branch close to the ground develops roots from which a new tree grows.

Levee: High bank of sediment deposited by a very silty river.

Lichens: Plantlike organisms that are combinations of algae and fungi. The algae produces the food for both by means of photosynthesis.

Limnetic zone: The deeper, central region of a lake or pond where no plants grow.

Littoral zone: The area along the shoreline that is exposed to the air during low tide; also called intertidal zone.

Longshore currents: Currents that move along a shoreline.

Lowland rain forest: Rain forest found at elevations up to 3,000 feet (900 meters).

Low tide: A lowering of the surface level of the ocean caused by Earth's rotation and the gravitational pull of the sun and moon.

Macrophytic: Term used to describe a large plant.

Magma: Molten rock from beneath Earth's crust.

Mangrove swamp: A coastal saltwater swamp found in tropical and subtropical areas.

Marine: Having to do with the oceans.

Marsh: A wetland characterized by poorly drained soil and by plant life dominated by nonwoody plants.

Mature stream: A stream with a moderately wide channel and sloping banks.

Meandering stream: A stream that winds snakelike through flat countryside.

Mesopelagic zone: An oceanic zone based on depth that ranges from 650 to 3,300 feet (200 to 1,000 meters).

Mesophytes: Plants that live in soil that is moist but not saturated.

Mesophytic: Term used to describe a forest that grows where only a moderate amount of water is available.

Mid-ocean ridge: A long chain of mountains that lies under the World Ocean.

Migratory: Term used to describe animals that move regularly from one place to another in search of food or to breed.

Mixed-grass prairie: North American grassland with a variety of grass species of medium height.

Montane rain forest: Mountain rain forest found at elevations between 3,000 and 10,500 feet (900 and 3,200 meters).

Mountain blanket bogs: Blanket bogs in Ireland that are more than 656 feet (200 meters) above sea level.

Mouth: The point at which a river or stream empties into another river, a lake, or an ocean.

Muck: A type of gluelike bog soil formed when fully decomposed plants and animals mix with wet sediments.

Muskeg: A type of wetland containing thick layers of decaying plant matter.

Mycorrhiza: A type of fungi that surrounds the roots of conifers, helping them absorb nutrients from the soil.

Neap tides: High tides that are lower and low tides that are higher than normal when the Earth, sun, and moon form a right angle.

Nekton: Animals that can move through the water without the help of currents or wave action.

Neritic zone: That portion of the ocean that lies over the continental shelves.

Nomads: People or animals who have no permanent home but travel within a well-defined territory determined by the season or food supply.

North Atlantic Drift: A warm ocean current off the coast of northern Scandinavia.

Nutrient cycle: Natural cycle in which mineral nutrients are absorbed from the soil by tree roots and returned to the soil when the tree dies and the roots decay.

Oasis: A fertile area in the desert having a water supply that enables trees and other plants to grow there.

Ocean: The large body (or bodies) of saltwater that covers more than 70 percent of Earth's surface.

Oceanography: The exploration and scientific study of the oceans.

Old stream: A stream with a very wide channel and banks that are nearly flat.

Omnivore: Organism that eats both plants and animals.

Ooze: Sediment formed from the dead tissues and waste products of marine plants and animals.

Oxbow lake: A curved lake formed when a river abandons one of its bends.

Oxygen cycle: Natural cycle in which the oxygen taken from the air by plants and animals is returned to the air by plants during photosynthesis.

Pack ice: A mass of large pieces of floating ice that have come together on an open ocean.

Pamir: A high altitude grassland in Central Asia.

Pampa: A tropical grassland found in South America.

Pampero: A strong, cold wind that blows down from the Andes Mountains and across the South American pampa.

Pantanal: A wet savanna that runs along the Upper Paraguay River in Brazil.

Parasite: An organism that depends upon another organism for its food or other needs.

Passive margin: A continental margin free of earthquake or volcanic action, and in which few changes take place.

Peat: A type of soil formed from slightly decomposed plants and animals.

Peatland: Wetlands characterized by a type of soil called peat.

Pelagic zone: The water column of the ocean.

Perennials: Plants that live at least two years or seasons, often appearing to die but returning to "life" when conditions improve.

Permafrost: Permanently frozen topsoil found in northern regions.

Permanent stream: A stream that flows continually, even during a long, dry season.

Pesticides: Poisons used to kill anything that is unwanted and is considered a pest.

Phosphate: An organic compound used in making fertilizers, chemicals, and other commercial products.

Photosynthesis: The process by which plants use the energy from sunlight to change water (from the soil) and carbon dioxide (from the air) into the sugars and starches they use for food.

Phytoplankton: Tiny, one-celled algae that float on ocean currents.

Pingos: Small hills formed when groundwater freezes.

Pioneer trees: The first trees to appear during primary succession; they include birch, pine, poplar, and aspen.

Plankton: Plants or animals that float freely in ocean water; from the Greek word meaning "wanderer."

Plunge pool: Pool created where a waterfall has gouged out a deep basin.

Pocosin: An upland swamp whose only source of water is rain.

Polar climate: A climate with an average temperature of not more than 50°F (10°C) in July.

Polar easterlies: Winds occurring near Earth's poles that blow in an easterly direction.

Pollination: The carrying of pollen from the male reproductive part of a plant to the female reproductive part of a plant so that reproduction may occur.

Polygons: Cracks formed when the ground freezes and contracts.

Pond: A body of inland water that is usually small and shallow and has a uniform temperature throughout.

Pool: A deep, still section in a river or stream.

Prairie: A North American grassland that is defined by the stature of grass groups contained within (tall, medium, and short).

Prairie potholes: Small marshes no more than a few feet deep.

Precipitation: Rain, sleet, or snow.

Predator: An animal that kills and eat other animals.

Primary succession: Period of plant growth that begins when nothing covers the land except bare sand or soil.

Producers: Plants and other organisms in the food web of a biome that are able to make food from nonliving materials, such as the energy from sunlight.

Profundal zone: The zone in a lake where no more than 1 percent of sunlight can penetrate.

Puna: A high-altitude grassland in the Andes Mountains of South America.

Rain forest: A tropical forest, or jungle, with a warm, wet climate that supports year-round tree growth.

Raised bogs: Bogs that grow upward and are higher than the surrounding area.

Rapids: Fast-moving water created when softer rock has been eroded to create many short drops in the channel; also called white water.

Reef: A ridge or wall of rock or coral lying close to the surface of the ocean just off shore.

Rhizomes: Plant stems that spread out underground and grow into a new plant that breaks above the surface of the soil or water.

Rice paddy: A flooded field in which rice is grown.

Riffle: A stretch of rapid, shallow, or choppy water usually caused by an obstruction, such as a large rock.

Rill: Tiny gully caused by flowing water.

Riparian marsh: Marsh usually found along rivers and streams.

Rip currents: Strong, dangerous currents caused when normal currents moving toward shore are deflected away from it through a narrow channel; also called riptides.

River: A natural flow of running water that follows a well-defined, permanent path, usually within a valley.

River system: A river and all its tributaries.

Salinity: The measure of salts in ocean water.

Salt lake: A lake that contains more than 0.1 ounce of salt per quart (3 grams per liter) of water.

Salt pan: The crust of salt left behind when a salt or soda lake or pond dries up.

Saltwater marsh: A wetland fed by saltwater and characterized by poorly drained soil and plant life dominated by nonwoody plants.

Saltwater swamp: A wetland fed by saltwater and characterized by poorly drained soil and plant life dominated by trees.

Saturated: Soaked with water.

Savanna: A grassland found in tropical or subtropical areas, having scattered trees and seasonal rains.

Scavenger: An animal that eats decaying matter.

School: Large gathering of fish.

Sea: A body of saltwater smaller and shallower than an ocean but connected to it by means of a channel; sea is often used interchangeably with ocean.

Seafloor: The ocean basins; the area covered by ocean water.

Sea level: The height of the surface of the sea. It is used as a standard in measuring the heights and depths of other locations such as mountains and oceans.

Seamounts: Isolated volcanoes on the ocean floor that do not break the surface of the ocean.

Seashore: The strip of land along the edge of an ocean.

Secondary succession: Period of plant growth occurring after the land has been stripped of trees.

Sediments: Small, solid particles of rock, minerals, or decaying matter carried by wind or water.

Seiche: A wave that forms during an earthquake or when a persistent wind pushes the water toward the downwind end of a lake.

Seif dunes: Sand dunes that form ridges lying parallel to the wind; also called longitudinal dunes.

Shelf reef: A type of coral reef that forms on a continental shelf having a hard, rocky bottom. A shallow body of water called a lagoon may be located between the reef and the shore.

Shoals: Areas where enough sediments have accumulated in the river channel that the water is very shallow and dangerous for navigation.

Shortgrass prairie: North American grassland on which short grasses grow.

Smokers: Jets of hot water expelled from clefts in volcanic rock in the deep-seafloor.

Soda lake: A lake that contains more than 0.1 ounce of soda per quart (3 grams per liter) of water.

Softwoods: Woods usually produced by coniferous trees.

Sonar: The use of sound waves to detect objects.

Soredia: Algae cells with a few strands of fungus around them.

Source: The origin of a stream or river.

Spit: A long narrow point of deposited sand, mud, or gravel that extends into the water.

Spores: Single plant cells that have the ability to grow into a new organism.

Sport fishing: Fishing done for recreation.

Spring tides: High tides that are higher and low tides that are lower than normal because the Earth, sun, and moon are in line with one another.

Stagnant: Term used to describe water that is unmoving and contains little oxygen.

Star-shaped dunes: Dunes created when the wind comes from many directions; also called stellar dunes.

Steppe: A temperate grassland found mostly in southeast Europe and Asia.

Stone circles: Piles of rocks moved into a circular pattern by the expansion of freezing water.

Straight stream: A stream that flows in a straight line.

Strait: The shallow, narrow channel that connects a smaller body of water to an ocean.

Stream: A natural flow of running water that follows a temporary path that is not necessarily within a valley; also called brook or creek. Scientists often use the term to mean any natural flow of water, including rivers.

Subalpine forest: Mountain forest that begins below the snow line.

Subduction zone: Area where pressure forces the seafloor down and under the continental margin, often causing the formation of a deep ocean trench.

Sublittoral zone: The seashore's lower zone, which is underwater at all times, even during low tide.

Submergent plant: A plant that grows entirely beneath the water.

Subpolar gyre: The system of currents resulting from winds occurring near the poles of Earth.

Subsistence fishing: Fishing done to obtain food for a family or a community.

Subtropical: Term used to describe areas bordering the equator in which the weather is usually warm.

Subtropical gyre: The system of currents resulting from winds occurring in subtropical areas.

Succession: The process by which one type of plant or tree is gradually replaced by others.

Succulents: Plants that appear thick and fleshy because of stored water.

Sunlit zone: The uppermost part of the ocean that is exposed to light; it reaches down to about 650 feet (200 meters) deep.

Supralittoral zone: The seashore's upper zone, which is never underwater, although it may be frequently sprayed by breaking waves; also called the splash zone.

Swamp: A wetland characterized by poorly drained soil, stagnant water, and plant life dominated by trees.

Sward: Fine grasses that cover the soil.

Swell: Surface waves that have traveled for long distances and become more regular in appearance and direction.

Taiga: Coniferous forest found in areas bordering the Arctic tundra; also called boreal forest.

Tallgrass prairie: North American grassland on which only tall grass species grow.

Tannins: Chemical substances found in the bark, roots, seeds, and leaves of many plants and used to soften leather.

Tectonic action: Movement of Earth's crust, as during an earthquake.

Temperate bog: Peatland found in temperate climates.

Temperate climate: Climate in which summers are hot and winters are cold, but temperatures are seldom extreme.

Temperate zone: Areas in which summers are hot and winters are cold but temperatures are seldom extreme.

Thermal pollution: Pollution created when heated water is dumped into the ocean. As a result, animals and plants that require cool water are killed.

Thermocline: Area of the ocean's water column, beginning at about 1,000 feet (300 meters), in which the temperature changes very slowly.

Thermokarst: Shallow lakes in the Arctic tundra formed by melting permafrost; also called thaw lakes.

Tidal bore: A surge of ocean water caused when ridges of sand direct the ocean's flow into a narrow river channel, sometimes as a single wave.

Tidepools: Pools of water that form on a rocky shoreline during high tide and that remain after the tide has receded.

Tides: Rhythmic movements, caused by Earth's rotation and the gravitational pull of the sun and moon, that raise or lower the surface level of the oceans.

Tombolo: A bar of sand that has formed between the beach and an island, linking them together.

Trade winds: Winds occurring both north and south of the equator to about 30 degrees latitude; they blow primarily east.

Transverse dunes: Sand dunes lying at right angles to the direction of the wind.

Tree: A large woody perennial plant with a single stem, or trunk, and many branches.

Tree line: The elevation above which trees cannot grow.

Tributary: A river or stream that flows into another river or stream.

Tropical: Term used to describe areas close to the equator in which the weather is always warm.

Tropical tree bog: Bog found in tropical climates in which peat is formed from decaying trees.

Tropic of Cancer: A line of latitude about 25 degrees north of the equator.

Tropic of Capricorn: A line of latitude about 25 degrees south of the equator.

Tsunami: A huge wave or upwelling of water caused by undersea earthquakes that grows to great heights as it approaches shore.

Tundra: A cold, dry, windy region where trees cannot grow.

Turbidity current: A strong downward-moving current along the continental margin caused by earthquakes or the settling of sediments.

Turbine: An energy-producing engine.

Tussocks: Small clumps of vegetation found in marshy tundra areas.

Typhoon: A violent tropical storm that begins over the ocean.

Understory: A layer of shorter, shade-tolerant trees that grow under the forest canopy.

Upstream: The direction from which a river or stream is flowing.

Upwelling: The rising of water or molten rock from one level to another.

Veld: Temperate grassland in South Africa.

Venom: Poison produced by animals such as certain snakes and spiders.

Vertebrates: Animals with a backbone.

Wadi: The dry bed of a stream that flows only after rain; also called a wash or an *arroyo*.

Warm-bodied fish: Fish that can maintain a certain body temperature by means of a special circulatory system.

Wash: The dry bed of stream that flows only after rain; also called an *arroyo* or a *wadi*.

Water column: All of the waters of the ocean, exclusive of the sea bed or other landforms.

Water cycle: Natural cycle in which trees help prevent water runoff, absorb water through their roots, and release moisture into the atmosphere through their leaves.

Waterfall: A cascade of water created when a river or stream falls over a cliff or erodes its channel to such an extent that a steep drop occurs.

Water table: The level of groundwater.

Waves: Rhythmic rising and falling movements in the water.

Westerlies: Winds occurring between 30 degrees and 60 degrees latitude; they blow in a westerly direction.

Wet/dry cycle: A period during which wetland soil is wet or flooded followed by a period during which the soil is dry.

Wetlands: Areas that are covered or soaked by ground or surface water often enough and long enough to support plants adapted for life under those conditions.

Wet meadows: Freshwater marshes that frequently dry up.

Xeriscaping: Landscaping method that uses drought tolerant plants and efficient watering techniques.

Xerophytes: Plants adapted to life in dry habitats or in areas like salt marshes or bogs.

Young stream: A stream close to its headwaters that has a narrow channel with steep banks.

Zooplankton: Animals, such as jellyfish, corals, and sea anemones, that float freely in ocean water.

Grassland

A grassland is a biome in which the dominant plants are grasses rather than trees or tall shrubs. Often described as "seas of grass," grasslands cover about one fourth of Earth's surface. They are usually found in the interiors of every continent except Antarctica.

Although grasslands vary in climate and the type of plant and animal life they support, most have several things in common. They are covered with grasses, which may be of different heights and varieties. They are usually windy and dry for part of the year. They occur primarily on flat land or gently rolling hills, but a few are found on mountains where the environment is suitable. Grasslands are considered transition zones between deserts, which receive little rain, and forests, which get a lot of rain.

How Grasslands Develop

Grasslands develop as a result of changes in climate, changes in plant communities, and fires.

Climatic change   Grasslands first appeared millions of years ago after mountains formed and caused climates to change. In North America, for example, the Rocky Mountains blocked moist air traveling across the continent from the Pacific Ocean, making the middle part of the continent drier. This caused trees to die and be replaced by grasses, which could adapt to the drier climate. The same process happened on other continents, allowing grasslands to form in places such as central Asia and South America. Grasslands throughout the world were fairly well established about 5 million years ago, covering more than 40 percent of Earth's surface.

Succession   Grasslands also develop through a process called succession, a slow sequence of changes in a plant community. In dry areas, the growth of mosses and lichens (LY-kens) may be followed by the growth of leafy,

WORDS TO KNOW

Artificial grassland: A grassland created by humans.

Carnivore: A meat-eating plant or animal.

Estivation: An inactive period experienced by some animals during very hot months.

Herbivore: An animal that eats only plant matter.

Pampa: A tropical grassland found in South America.

Pantanal: A wet savanna that runs along the Upper Paraguay River in Brazil.

Puna: A high-altitude grassland in the Andes Montains of South America.

Scavenger: An animal that eats decaying matter.

Steppe: A temperate grassland found mostly in southeast Europe and Asia.

Veld: Temperate grassland in South Africa.

nonwoody plants. Gradually, the grasses, which are hardier plants, begin to take over and become the major form of plant life in the area.

In ponds or other areas of still or slow-moving water, submerged plants like pondweed, grow beneath the water. Dead stems and leaves from these plants make the water thick, shallow, and slow moving. This dead matter forms a thick layer of organic material in which plants that must be anchored in soil, such as reeds and grasses, begin to grow. As this process continues, the pond fills with decaying plants until the water is gone and a grassland has developed.

Fires   Fires help form and sustain grasslands. Most naturally occurring fires are started by lightning. Lightning strikes the ground, igniting dried grass. All vegetation (trees, shrubs, flowers) and grass are completely burned. The grass regrows because it has adapted to this dry environment and has a very deep root system and, often, underground stems. Fire helps to eliminate competition from trees, shrubs, and flowers for nutrients, water, and growing space, making it easier for grass to grow.

People may start fires that help establish grasslands. As far back as the Stone Age, hunters burned forests so that grasses would grow and attract the wild animals they relied on for food. Later, shepherds and herders burned brush and trees to encourage the growth of grasses for grazing.

Kinds of Grasslands

Grasslands may be classified as tropical, temperate, and upland grassland.

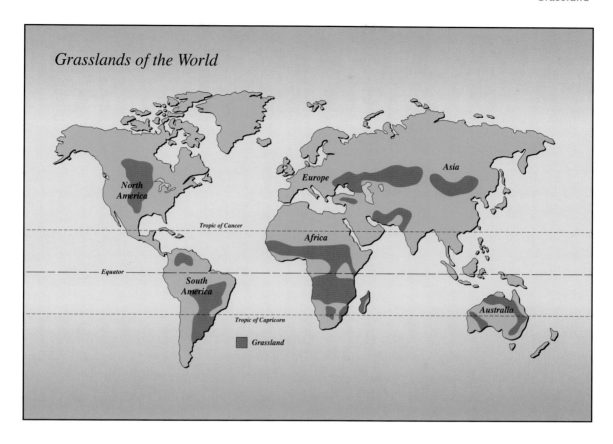

Grasslands of the World

Tropical grassland   Tropical grasslands, called savannas when they have scattered trees, are found in regions around the equator. Savannas are hot all year and have long dry seasons followed by very wet seasons. Savanna grasses vary across continents, depending on average rainfall. Thorn trees, for example, are found in North African savannas and can range from 13 to 50 feet (4 to 15 meters) in height. More trees grow in savannas than in any other type of grassland.

In Africa, Australia, South America, and India, savannas are transition zones between the rain forests and the deserts of the higher latitudes (distances north or south of the equator). In South America, the Llanos (YAH-nos) covers parts of northwestern Venezuela and northeastern Colombia, while larger and more wooded grasslands lie across southeastern Venezuela and southern Guyana. Brazilian grasslands include the Cerrado, a highland region, and the Pantanal, a wet savanna within the Cerrado. The Pantanal is the world's largest area of continental wetlands.

Carefully controlled fires in grasslands eliminates the competition from trees and other plants, making it easier for the grass to grow. IMAGE COPYRIGHT PHOTOSKY 4T COM, 2007. USED UNDER LICENSE FROM SHUTTERSTOCK.COM.

Almost half of Africa, about 5,000,000 square miles (13,000,000 square kilometers), is covered with savanna, making it the world's largest tropical grassland. A well-known savanna in Tanzania is the Serengeti (sur-in-GEHT-ee) Plain. Savannas are also found in Australia and on India's Deccan Plateau.

Temperate grassland   Temperate grasslands are those with moderate climates located farther from the equator. They have fewer trees than tropical grasslands and the grass is shorter.

Most savannas are tropical, but some occur in places with temperate climates, where they form borders between other grasslands and deciduous (trees that lose their leaves at the end of the growing season) forests. For example, oak savannas, 20 percent of which are tree covered, are found in midwestern states of the United States, such as Illinois, Indiana, Missouri, Kentucky, and Wisconsin. The major temperate grasslands include prairies, steppes, pampas, and velds.

Prairie   The vast sweeping grasslands in North America are generally called prairies. The many different types of prairie grasses grow in a variety of textures and colors, which range from green to gold. These grasses vary in height but typically are 5 feet (1.5 meters) tall.

Three prairie regions are located east of the Rocky Mountains. The shortgrass prairie begins in the foothills of the Rockies and stretches east,

Big bluestem tallgrass prairie growing in the Homestead National Monument of America, Beatrice, Nebraska. IMAGE COPYRIGHT WELDON SCHLONEGER, 2007. USED UNDER LICENSE FROM SHUTTERSTOCK.COM.

where it merges with the mid-grass region of the Great Plains. The tallgrass prairie is easternmost, ending in Illinois.

Other prairie regions include the desert grassland of the southwestern United States, the intermountain grassland in the northwestern United States, and the central valley grassland in California. Within these prairie regions are subprairies—gravel prairies, hill prairies, sand prairies, and dry sand prairies. Gravel prairies grow on gravelly soil, while hill prairies are found in clearings on forested slopes. Sand prairies are characterized by sandy soils, and dry sand prairies are found on the crests of sand dunes.

Steppe   *Steppe* is the Russian word for "grassy plain." A steppe is similar to a prairie, except that the grasses are shorter. Steppes are found primarily in southeastern Europe and Asia. One of the world's largest grasslands, the Eurasian steppe, stretches thousands of miles from Europe to China. Siberia is also home to steppes.

In general, there are two kinds of steppe: the meadow steppes of the North and the dry steppes of the South.

Pampas   The pampas, which is Spanish for "plains," are temperate grasslands in South America. The land is flat and rolling and supports about twenty species of grasses that are tall and reedlike with silvery plumelike flowers. Many pampas grasses grow in tussocks (small clumps).

Veld   In South Africa a temperate grassland is called a veld (FELT) or veldt, the Dutch word for "field." Velds are found at different elevations—

high, middle, and low. They are home to a large variety of vegetation. The highveld, for example, is dominated by red grasses that may be sweet or sour. The sweet grasses provide a good source of food for animals.

Upland grassland   A few grasslands exist at high altitudes where it is too cold for trees to grow and only the hardiest of grasses survive. Upland grasslands include alpine meadows and alpine savannas. Alpine meadows are mountainous grasslands and occur in mountain ranges throughout the world. Alpine savannas occur in the tropics (the part of Earth's surface between the Tropic of Capricorn and the Tropic of Cancer, two lines of latitude above and below the equator). One example of alpine meadows is the *Puna* of the Andes Mountains of southeastern Peru and western Bolivia. Native and introduced grasses grow at elevations ranging from about 9,000 to 11,000 feet (2,800 to 3,850 meters). Forage grasses, on which grazing animals feed, grow at lower elevations.

Climate

Climate, especially rainfall and wind conditions, is the most important factor in grassland survival. Grasslands located deep in the interiors of continents in the Northern Hemisphere face extreme weather from tornadoes, droughts

Alpine meadows often have colorful flowers growing with the green grasses. IMAGE COPYRIGHT IGOR SMICHKOV, 2007. USED UNDER LICENSE FROM SHUTTERSTOCK.COM.

(dry periods), blizzards, and dust storms. Temperate grasslands in the Southern Hemisphere, such as the pampas in Argentina, are closer to the moderating effect of the oceans, which makes their climates less extreme.

Temperature Since tropical grasslands are near the equator, their climate is warm all year. The winter is dry and the summer brings a short, but very wet, rainy season. The average summer temperature on a tropical savanna is above 80°F (26.6°C). Winter averages about 65°F (18°C).

Temperatures in temperate grasslands vary according to how far north or south of the equator they are located and how far inland from the oceans. In general, summers are hot and winters are cold. On the North American prairies, summer temperatures often reach 100°F (37.7°C). Prairies in Canada, which are farther from the equator, can be quite cold in winter, with the temperature often sinking to 14°F (-10°C).

The scarlet ibis lives on the Llanos, the tropical savanna found in Venezuela and northeastern Colombia. IMAGE COPYRIGHT RACHEAL GRAZIAS, 2007. USED UNDER LICENSE FROM SHUTTERSTOCK.COM.

Precipitation Precipitation (rain, snow, or sleet) is a key factor in determining the nature of a grassland, especially its soil and the type of plant life growing there. Grasslands typically receive 10 to 30 inches (25 to 76 centimeters) of rain every year; however, periods of drought are common. If rainfall decreases significantly over many years, grassland will become a desert. Likewise, an increase in rainfall over a long period of time encourages the growth of forests. Grasslands closer to the equator tend to have longer rainy seasons and a moister climate than grasslands farther from the equator.

Tropical savannas have what is called a monsoon climate, which means winds blow from different directions during different seasons. Winters are dry, and the dry period may last five to seven months. Summers, when the wind shifts direction, bring periods of heavy rain. Typical annual precipitation ranges from 25 to 60 inches (64 to 152 centimeters), but amounts vary depending on location. Grasslands in Australia, for example, receive fewer than 20 inches (50 centimeters) of

Wind turbines are powered by the frequent winds that cross the prairies. IMAGE COPYRIGHT INAKI ANTOÑANA PLAZA, 2007. USED UNDER LICENSE FROM SHUTTERSTOCK.COM.

rainfall in the east and 30 inches (76 centimeters) in the west. Grasslands in South America receive as many as 60 inches (150 centimeters).

Temperate grasslands generally receive less precipitation than tropical grasslands. They average between 12 to 40 inches (31 to 102 centimeters) per year, with much of the rain falling in the late spring and early summer.

Dry periods in both types of grasslands are important to their survival. For example, dry conditions promote fires, which is nature's way of encouraging new growth.

Wind Without trees to block the wind, it sweeps across grasslands at a high speed. In tropical grasslands, winds bring the soaking monsoon rains. Wind also helps to dry things out again because it aids evaporation. This reduction in moisture helps keep trees from growing and taking over the grassland. Wind plays a major role in spreading fires during the dry season.

Some grassland winds are so ferocious and well known that they have been given names. The buran is a strong northeasterly wind that blows over the steppes of Russia and Siberia, bringing blizzards in winter and hot dust in summer. The pampero of Argentina is a strong, cold wind that sweeps in from the Andes Mountains across the South American pampas. The warm, dry winter wind that blows over the Rocky Mountains in North America is called the chinook.

Geography of Grasslands

The geography of grasslands involves landforms, elevation, soil, and water sources.

Landforms The dominant feature of most grasslands is flat terrain or low, rolling hills. In African savannas, small hills called kopjes (KOPP-ees) are formed from rocks. Kopjes have their own type of vegetation and wildlife. In a land with few trees, kopjes provide shelter, shade, and protection for animals. The North American prairies often contain potholes, grass-filled depressions in the ground that fill with water after heavy rains.

Elevation Grasslands are found at many elevations. The South African veld, for example, is divided into zones based on elevation. The lowveld is

Twister!

Grasslands often lie in the path of violent windstorms called tornadoes. In a tornado, a huge column of air spins like a top at speeds of up to 300 miles (480 kilometers) per hour. Some have been recorded as spinning up to 500 miles (800 kilometers) per hour. Tornadoes are often called funnel-clouds because of their funnel-like shape. They are also called twisters because they spin.

While tornadoes occur in many countries around the world, they are most common in the grassland areas of the United States, especially in Kansas and Oklahoma, where few trees block the wind. As many as 1,000 tornadoes strike there every year. They develop during thunderstorms that form when warm tropical air hits cooler northern air and when the humidity is high and the air near the ground is unstable. Tornadoes travel a narrow path, usually about 600 feet (200 meters) wide, at an average speed of 40 miles (64 kilometers) per hour. Inside the center of the funnel cloud is a very strong updraft (an air current that moves upward). The updraft is so powerful it can lift heavy objects, like trees, cars, and even homes, into the air. Tornadoes can leave a path of destruction up to 15 miles (24 kilometers) long, often causing millions of dollars in damage and killing many people.

found between 500 and 2,000 feet (152 and 610 meters), the middle veld between 2,000 and 4,000 feet (610 and 1,219 meters), and the highveld between 4,000 and 6,000 feet (1,219 and 1,828 meters).

Colder grasslands are found at much higher elevations. The altiplano of the central Andes Mountains in Bolivia and Peru is about 12,000 feet (3,650 meters) above sea level.

Soil   Grassland soil helps to determine what plants and animals can survive in an area.

Tropical savanna soils   Soils in savannas are called "red earth" and are mostly sandy and dusty. Their red color comes from a high iron content. The long dry periods between the rainy seasons prevents dead plant matter from decomposing (breaking down) and releasing nutrients into savanna soil. This soil is not as rich as in other grasslands. Termites and other burrowing creatures turn the soil, helping air to circulate and water to reach lower levels.

Temperate grassland soils   Temperate grassland soils are rich with humus, a dark, moist layer composed of decayed plant and animal matter and small grains of rock. Humus is rich in nutrients, such as nitrogen, that are vital to plant growth. It is spongy and able to store moisture.

The Role of Soil in the Food Web

Plants obtain some of their nutrients from the soil. These nutrients include minerals such as nitrogen and phosphorus, and acids such as carbonic and citric acids. Not all soils are the same and some contain a better balance of nutrients than others. Studies have shown that the quality of the soil in which a plant is grown can affect not only the plant itself, but also the creatures that eat the plant.

In a study at the University of Missouri, hay was grown on three plots of ground, all lacking nutrients. One plot (A) remained untreated. It produced about 1,700 pounds (3,740 kilograms) of hay per acre. Nitrogen fertilizer was added to the second plot (B), and it produced almost twice as much hay, about 3,200 pounds (7,040 kilograms) per acre. When the hay from both plots was fed to rabbits, the rabbits that ate hay from plot B did not grow as big as those fed hay from plot A. Although the nitrogen produced more plants per acre, those plants were not as nutritious for the rabbits.

A third plot (C) was given a balanced fertilizer to provide all the minerals the hay needed. This plot produced less hay than plot B, only 2,400 pounds (5,280 kilograms) per acre. However, when the rabbits ate it they grew twice as big as the rabbits that ate hay from plot B and 35 percent bigger than the rabbits that ate hay from plot A.

The three main types of temperate grassland soil are black earth, prairie soil, and chestnut soil.

Black earth In dry climates, there is not enough rain to wash the humus down into the soil, so it remains close to the top, producing "black earth." Called *chernozem* in Russian, black earth is an extremely fertile soil that provides excellent nutrition for crops like wheat and soybeans. Black earth is found on the steppes of Russia and central Asia, on the pampas of South America, in Australia, and on some North American prairies.

Prairie soil In moist climates with an average rainfall of 25 to 40 inches (64 to 102 centimeters), rain pushes humus deeper into the ground, producing brown prairie soils. These very fertile soils cover parts of eastern Europe and what is known as the Corn Belt in the United States. This belt runs through parts of Ohio, Indiana, Illinois, Iowa, Minnesota, South Dakota, Nebraska, Kansas, and Missouri. Prairie soils may be wet or dry and contain varying amounts of sand or gravel.

Chestnut soil Chestnut soils are found in the driest of the temperate grasslands, on parts of the South African veld, in the Argentine pampas, and on the high plains east of the Rocky Mountains. These soils are light or dark brown depending on their humus content, which makes them more or less fertile.

Water sources   Rainy seasons in tropical grasslands create waterholes, swell rivers, and fill flood plains. Some tropical grasslands, like the Pantanal in Brazil, get so drenched they are under water for part of the year.

Ponds, lakes, streams, rivers, and marshes are found throughout temperate grasslands.

Plant Life

Grasslands support a wide variety of plant life, including more than 10,000 species of grasses worldwide. Grassland plants include algae (AL-jee), fungi (FUN-ji), lichens, and green plants.

Algae, fungi, and lichens   It is generally recognized that algae, fungi, and lichens do not fit neatly into the plant or animal categories.

Algae   Most algae are one-celled organisms too small to be seen by the naked eye. They make their food through photosynthesis (foh-toh-SIHN-thuh-sihs), the process by which plants use energy from the sun to change water and carbon dioxide from the air into sugars and starches they require for growth. Blue-green algae, also called blue-green bacteria, grow close to the ground in moist areas, often in mud after a rain. These algae help transform nitrogen in the soil so it can be absorbed by plants.

Fungi   Fungi are plantlike organisms that cannot make their own food by means of photosynthesis. Instead, they grow on decaying organic matter or live as parasites (organisms that depend upon another organism for survival). Hundreds of species of fungi live in grassland soil and grow best in a damp environment. In the African savanna, for example, the termite-mound fungus grows on moist termite droppings. The fungi breaks down food matter in the droppings that the termites could not digest. This food matter is then re-eaten by the termites.

Lichens   Lichens are combinations of algae and fungi. The alga produces food for both through photosynthesis. It is believed that the fungus protects the algae from dry conditions. Lichens found in grasslands grow close to the ground in moist environments.

Green plants   Most green plants need several basic things to grow: light, air, water, warmth, and nutrients. In a grassland, more than half of all plant tissue—roots, stems, and rhizomes—lies underground. It is here that the plants absorb their nutrients, water, and oxygen. Savanna elephant grasses, for example, have roots that reach 10 feet (3 meters) deep.

Tough as Grass

Grass may not be as hard as nails, but it can be just as tough. It can survive drought, subzero temperatures, high winds, fires, mowing, and being trampled upon. Grass is able to grow again even after being severely damaged. Growing points on its stems and buds close to the ground can start new growth.

A tall, rough grass called slough (SLOO) grass grows along the edges of prairies close to water sources. Slough grass reaches about 9 feet (2.7 meters) in height. Nicknamed "ripgut," its stiff, razor-like leaves can cut a person's hands or an animal's mouth. While slough grass is not preferred for eating, its tough qualities make it useful for roof thatching.

Common green plants Grasses, sedges, and forbs are the most common green plants found on grasslands. A few scattered trees and shrubs also grow.

Grasses have round, hollow stems and long narrow leaves called blades. The blades grow from the base of the plant so that, when the grass is cut off at the top, it continues to grow. More than half of the grass tissue is underground, which helps the plants survive harsh weather. Some grasses, such as needle-and-thread grass, have adapted to prefer cooler rather than warmer climates.

Grasses are categorized primarily according to their height. The tallest grasses grow in tropical areas where rainfall is greater. Bamboo, one of the tallest and strongest grasses in the world, can be 180 feet (60 meters) tall. Bamboo grows quickly, sometimes at the rate of 3 feet (0.9 meter) a day. It is found throughout the tropics. A shorter species of bamboo grows in the southern United States.

Mixed-height grasses with flowering stems grow in temperate regions. On the North American prairies the grasses are categorized as shortgrass, mixed-grass, and tallgrass. Shortgrasses, which include grama grass and buffalo grass, grow to about 18 inches (45 centimeters) in height. Little bluestem, needlegrass, and foxtail barley, three mixed-grasses, grow to 3 feet (90 centimeters) tall. Tallgrasses, ranging from 3.3 to 10 feet (1 to 3 meters), include big bluestem, slough (SLOO) grass, cordgrass, and Indian grass.

Shorter grasses that grow in arid lowlands like the dry, southern Serengeti Plain, need less rainfall to survive. These include Rhodes grass and red oat grass. Most are less than 8 inches (20 centimeters) high.

Sedges are perennial, flowering herbs that closely resemble grasses. Found all over the world, they range in size from 0.4 inch to 16 feet (1 centimeter to 5 meters). They have solid, usually triangular, stems and grasslike green leaves arranged in three rows. Sedges prefer a damper environment than grasses and are often found on the edges of marshes, ponds, or other watery locations. The remote sedge, carnation sedge, and yellow nut sedge are common species.

African landscape with a beautiful Acacia tree, Hwange National Park, Zimbabwe, southern Africa. IMAGE COPYRIGHT ECOPRINT, 2007. USED UNDER LICENSE FROM SHUTTERSTOCK.COM.

Forbs are flowering, broad-leaved plants without the woody stems of trees and shrubs. Wildflowers are prairie forbs and include blazing stars, sunflowers, purple coneflowers, bee balms, and shooting stars. Other North American forbs include gentian, milkweed, and fleabane. Forbs growing on the Russian steppes include anemones, red tulips, irises, and peonies.

Some temperate grasslands in drier regions support shrubs such as the mesquite and creosote bush. Along the wet edges of these grasslands, cottonwood, ash, and box elder can take root.

Baobabs and acacia trees thrive in the tropical savannas. Found in Africa and the Australian outback, the baobab has roots that go deep into the ground in search of water. It stores water in its thick trunk, which acts like a reservoir supplying water during the dry season. The African acacia produces as many as 20,000 seeds every year, ensuring that some will survive dry weather and grazing animals. As added protection for its survival, the acacia, known as a wattle in Australia, has long spines (like thorns) that are actually modified leaves. These spines keep grazing animals away and act as protection against drought, since they offer less surface area than normal leaves for moisture to evaporate.

Growing season   The length of a grassland's growing season varies depending on precipitation and temperature. Growing seasons in tropical grasslands start when the rains come and end when the rains end.

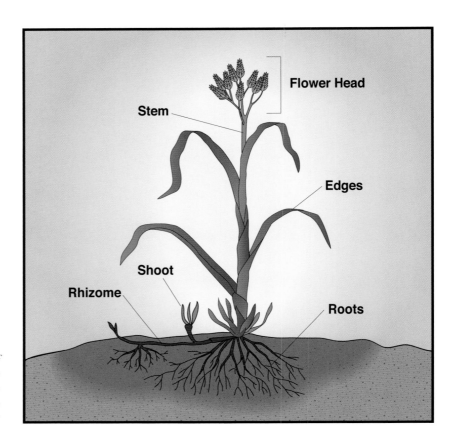

Illustration showing the parts of mixed-height grass. Since the blades of the grass grow from the base, the grass will continue to grow even after its top is cut off.

In temperate climates, the growing season lasts from 150 to 270 days and begins when the average temperature reaches about 50°F (10°C).

Reproduction   Green plants reproduce by several methods. One is pollination, in which the pollen from the male reproductive part of a plant, called the stamen in flowering plants, is carried by wind or insects to the female reproductive part of a plant, called a pistil in flowering plants. Pollen-gathering insects distribute pollen at different times of the day according to when a plant's flower is open. Lilies, for example, are closed in the morning and evening, but open during midday when the weather is warmer and insects are more active.

Many species of perennial prairie grasses (those that live for two or more growing seasons) reproduce with the help of rhizomes, modified stems that spread out under the soil and form new plants. Rhizomes grow roots and produce leaves, stems, and flowers that grow upward and out of the soil.

Annual grasses (those that live only one year) produce seeds with thick outer shells. These shells protect the seeds, which are spread by the wind or by attaching themselves to passing animals.

Endangered species   As grasslands become overrun by humans and animals, plants are endangered. Forbs are especially sensitive to abuse. When land is cultivated for crops or used for grazing animals, all of the native species in that area are destroyed.

Two endangered plants on the tallgrass prairies in North America are the leafy prairie clover and the prairie white-fringed orchid. The orchid is endangered because people dig it up to transplant into their home gardens and because pesticides accidentally kill the sphinx moth, an insect that pollinates the orchid.

Animal Life

Grassland animals range from very small aphids and worms to large African elephants. Different continents are home to different species.

Microorganisms   Microorganisms are tiny animals that cannot be seen by the human eye except with the help of a microscope. Those found in grasslands are mainly protozoa and bacteria, which live in the root systems of grasses and help with decomposition. Some protozoa live in the intestines of termites and make it possible for the termite to digest the wood fibers they eat.

Invertebrates   Invertebrates are animals that do not have a backbone. The invertebrate population of grasslands consists primarily of insects, such as dung (scarab) beetles, monarch butterflies, moths, and dragonflies.

Grubs, termites, and worms are invertebrates that play a key role in soil development. They churn the soil, allowing more oxygen and water to enter it. Worms rebuild the soil by digesting organic material and depositing this rich fertilizer in the ground.

Common grassland invertebrates   Two common grassland invertebrates are termites and grasshoppers. Termites are insects that resemble ants. Most of the more than 2,000 species live in the tropics, although some are found in temperate areas. Termites live by the millions in highly social and organized communities in underground nests or in mounds having many chambers. Each termite has a role in the colony and is ranked as either royalty, nobility, soldier, or worker.

The Web of Life: Protecting House and Home

Certain species of African ants make their homes in the bullhorn acacia tree. Animals that might be a threat to the tree are driven away by the ants, which bite and sting. The presence of the ants stimulates the tree to form galls, a kind of tumor that grows on the branches. The galls have spikes that help keep hungry animals away from both the tree and the ants.

Grasshoppers are one of the most common invertebrates on the grasslands. IMAGE COPYRIGHT BRUCE MACQUEEN, 2007. USED UNDER LICENSE FROM SHUTTERSTOCK.COM.

On the African savanna, termites build huge brown mounds out of soil cemented with saliva or droppings, making the mounds rock hard. Termite mounds can be more than 25 feet (7.6 meters) high, 98 feet (30 meters) wide, and may last for decades.

Grasshoppers are one of the most common grassland insects. They are plant eaters with powerful hind legs that allow them to jump and wings that allow them to fly. All grasslands around the world face periodic swarms (large populations) of grasshoppers. The species that swarm are called locusts. In 1889, one of the largest swarms ever seen formed over the Red Sea in Northeast Africa and covered an area about 2,000 miles (3,200 kilometers) long, darkening the sky for days. Locusts may eat every plant in their path, including the stalks. When they are finished, all crops, as well as plants and shrubs, are destroyed.

Food Caterpillars and grasshoppers are among the insects that eat leafy vegetation. Grubs eat grass roots, while some insects are scavengers, feeding on what is left behind by larger animals. Two types of scavengers are the carrion beetle and the flesh fly, both of which live on the African savanna. The African scarab beetle is also a scavenger, feeding on animal droppings.

Reproduction The first stage of an invertebrate's life cycle is spent as an egg. When the egg hatches, the emerging creature is called a larva. The larva stage is divided into several stages between which there is a shedding of the outer skin as the larva increases in size. During the third, or the pupal stage, the insect lives in yet another protective casing, like a cocoon. In the final stage, the adult emerges.

The reproductive cycle of the sphinx moth of the African savanna follows the wet and dry seasons. Eggs are laid on vegetation during the rainy season. By the time the dry season arrives, the eggs have already hatched into caterpillars that have been feeding on leaves. After several molts (stages

Sphinx moths can be found in grassy areas. IMAGE COPYRIGHT C.L. TRIPLETT, 2007. USED UNDER LICENSE FROM SHUTTERSTOCK.COM.

where insects increase in size and shed their outer skin) they burrow into the soil, where they remain in pupal form until it rains again and they emerge as adult moths. The scarab beetle lays its eggs inside a ball of animal droppings. The young beetles feed on the droppings after they have hatched.

Amphibians   Amphibians are vertebrates, animals with a backbone. Amphibians live at least part of their lives in water. Most are found in warm, moist regions and temperate zones where temperatures are seldom extreme.

Amphibians breathe through their skin, so they usually must remain close to water. Only moist skin can absorb oxygen, and if they are dry for too long they will die. On the African savanna, amphibians like the African bullfrog estivate (remain inactive) underground during the dry season. The termite frog estivates in a termite mound.

Common grassland amphibians   The green toad is common to the steppes. Once it has become an adult, it is able to live out of water. Since the steppe climate can be extreme, the toad hibernates underground when the weather is too hot or too cold.

The small North American cricket frog lives in grasslands near ponds or streams. These frogs have brown or green skin covered with bumps. They grow to about 1.2 inches (3 centimeters) in length.

Food   Amphibians eat insects and some small animals, using their long tongues to capture prey. Although they have teeth, amphibians do not chew but swallow their food whole. As larvae, they usually eat plants. Toads, for example, eat algae and water plants during the larval stage.

Reproduction   Mating and egg laying for most amphibians takes place in water. Male sperm are deposited in the water on top of eggs laid by the female. The African running frog lays its eggs in puddles that form when the rainy season begins. As the offspring develop into larvae and young adults, they have gills, which means they must live in water. Once they mature, they breathe through lungs and live on land.

Most amphibians reach adulthood at three or four years, breeding for the first time about one year after they become adults.

Reptiles   Reptiles are cold-blooded vertebrates that depend on the environment for warmth. They do not do well in extreme temperatures, either hot or cold. Reptiles are usually more active when temperatures become warmer. Many reptiles go through a period of hibernation in cold weather because they are so sensitive to the environment. Unlike amphibians, reptiles have waterproof skin, which allows them to move away from moist areas.

Common grassland reptiles   The boomslang snake is common throughout the African savannas south of the Sahara Desert. Up to 6.5 feet (2 meters) in length, the boomslang is shy and highly poisonous. Its fangs are set in the rear of its upper jaw, and a bite causes excessive bleeding and death. The boomslang eats lizards, frogs, and sometimes birds and rodents. Other snakes include the python and the coral snake.

The agama lizard is also found on the African savannas. The male has a blue-orange body and red head. These bright colors help it to be visible to females and competing males. Each male lizard has its own territory and mates with several females. Adults are around 12 inches (30 centimeters) long.

The western box turtle is a land turtle found in the prairies of North America. Box turtles grow to about 8 inches (20 centimeters) in length and have a high, round shell. They eat both plants and animals, including berries, mushrooms, insects, and worms.

Food   Most reptiles are carnivorous (meat-eating). For example, snakes consume their prey whole—and often alive—without chewing. It can often take an hour or more to swallow a large victim. Many snakes have fangs that are curved backward so their prey cannot escape.

Reproduction Most reptiles reproduce sexually, meaning the female egg is fertilized by the male sperm. Almost all species lay eggs. A rare few species bear live young.

Birds Hundreds of species of birds live on the grasslands. Many species must nest on the ground and perch on the grasses because there are so few trees.

Common grassland birds Representative of grassland birds are the ostrich, the prairie chicken, and the black hornbill.

The ostrich is the largest and fastest bird in the world, weighing around 350 pounds (159 kilograms) and standing about 8 feet (2.4 meters) tall. Although it has wings, it cannot fly, but it is able to run at speeds up to 45 miles (72 kilometers) per hour. The ostrich is found on grasslands in parts of Africa and Southwest Asia. Males mate with several females, each of which typically lays twelve very large eggs in a nest built on the ground. In this species, the male and female take turns sitting on the eggs until they hatch.

Formerly one of the most familiar grassland birds in the United States is the greater prairie chicken, a species that has become quite rare. Found on tallgrass prairies, these birds live on the ground and eat seeds. Females lay about twelve eggs that hatch in fewer than forty-two days. The chicks can fly by the time they are two weeks old. The lesser prairie chicken is more common and can be found on short-grass prairies.

Black hornbills are large birds that walk along the African savanna in search of insects. They range from 12 to 47 inches (30 to 120 centimeters) in length and have large heads, thin necks, broad wings, and long tails. Hornbills nest in the grooves of large trees. The male builds the nest around the female and seals her inside with the eggs, passing food to her through a small opening. The female breaks out of the nest after the eggs hatch.

Hitchhiking on the Grassland

The bustard is a large, heavy, longlegged grassland bird found in Europe, Asia, and Africa. When the male strolls around looking for food, a small, reddish bird called a bee-eater hitches a ride on his back. As the bustard moves, it stirs up insects that are nabbed and eaten by the bee-eater.

The burrowing owl is one of the few ground-dwelling owls, and lives in dry areas with low vegetation. IMAGE COPYRIGHT BOB BLANCHARD, 2007. USED UNDER LICENSE FROM SHUTTERSTOCK.COM.

An Upside-Down Nest

African weaverbirds build some of the most complex and unusual nests in the world. Their globe-shaped nest is woven from savanna grasses and has its entrance hole at the bottom. With its entrance hidden, the nest is protected from predatory birds that might ordinarily swoop down into the nest to steal eggs or baby birds. The nests of many weaverbirds are hung together in trees, forming a community of "apartments" that resembles the thatched roof of a house when seen from a distance. Parent birds use the same nest every year, and young birds add their nests to their parents' structure.

Food   Many grassland birds eat grains and seeds, which are readily available. Predatory (hunting) birds, like owls, hawks, and eagles, eat rodents and snakes, while others feed on insects. The green wood hoopoe eats butterflies or beetles. Tick birds, such as the groove-billed anis, eat ticks from the backs of large animals. Storks, vultures, and ravens are scavengers and dine on leftovers from a kill.

Reproduction   All birds reproduce by laying eggs. Usually, the males must attract the attention of the females. Therefore, most males are more brightly colored than the females, and some sing or perform dancing rituals. For example, the greater prairie chicken inflates orange air sacs near its throat when beginning its mating call.

After mating, female birds lay their eggs in nests made out of many different materials and built in a variety of places. Many species, like the lark and bobolink, nest on the ground. The parent bird, usually the female, sits on the eggs to keep them warm until they hatch.

Ostriches live in groups of five to fifty birds, and are often seen traveling with zebras or other grazing animals. IMAGE COPYRIGHT ECOPRINT, 2007. USED UNDER LICENSE FROM SHUTTERSTOCK.COM.

Mammals Mammals are warm-blooded vertebrates that are covered with at least some hair, bear live young, and produce their own milk. Hundreds of species of mammals live on grasslands around the world.

Grassland mammals Grassland mammals include prairie dogs and saigas. Saigas are endangered with a current population of about 40,000.

One of the most common North American prairie animals is the prairie dog. These small rodents weigh 1.5 to 3 pounds (0.7 to 1.4 kilograms) and live in prairie dog "towns" consisting of a network of underground tunnels. These tunnels provide a safe place to raise young, hibernate in winter, and hide from predators such as coyotes, snakes, and eagles. The many holes that lead into the tunnels can be found scattered over a wide area. The animals feed close by these holes and dart into them when danger is detected.

A male prairie dog mates with two or three females and protects the group, which lives in its own section of the town.

Getting the Most out of Every Meal

Many grasses are impossible for most mammals to digest. Some animals, like the bison, giraffe, and pronghorn antelope, are called ruminants and can eat tough grasses. These animals have a stomach with four chambers. After the grass is chewed and swallowed, it is stored in the first chamber, called the rumen, where microbes help break it down and soften it. Later, the animal brings the food back up into its mouth and chews it some more, a process called chewing cud. This cud is reswallowed and goes through all of the chambers of the stomach where it is completely digested. The four-chambered stomach probably helps grazing animals, which are always on the lookout for predators, eat large amounts very quickly and return to shelter where they can safely digest their meal.

Saigas are a type of antelope found on the steppes of Russia and Siberia. At only 2.5 feet (0.8 meter) tall, they must jump into the air to see above the grass when watching for predators. When threatened, they run away at speeds up to 50 miles (80 kilometers) per hour.

Saigas are plant-eaters that dine on forbs and grasses. Females often give birth to twins, which helps the survival of the species. Male saigas have a high mortality rate. Each year, more than half of the males starve to death.

Food Some mammals are herbivores (plant-eaters). Some herbivores, like the zebra, buffalo, wildebeest (WIHL-duh-beest), and antelope, are grazers. Grazers have front teeth that allow them to bite off grass close to the ground and back teeth to grind up the grass. Other herbivores, like giraffes and elephants, are browsers that nibble the leaves and bark of trees and shrubs.

Wildebeest and zebra graze in the Ngorongoro Crater in Tanzania. IMAGE COPYRIGHT VERA BOGAERTS, 2007. USED UNDER LICENSE FROM SHUTTERSTOCK.COM.

Other mammals, such as coyotes, lions, and cheetahs, are carnivores (meat-eaters). Jackals and hyenas are scavengers, meaning they eat the leftover carcasses and decaying meat from dead animals.

Reproduction   Mammals give birth to live young that develop inside the mother's body. Some mammals are helpless at birth, like the hare, while others, like the zebra, are able to walk and even run almost immediately. The mother nurses the young with her own milk until they are old enough to find their own food.

Endangered species   Many grassland animals are endangered because of overhunting, pollution, and the destruction of their habitats.

Before the end of the nineteenth century, there were as many as 400 million prairie dogs living in a single prairie dog town that stretched across North America. As grasslands were turned into farms, so many of these animals were killed that they faced extinction. Many national parks in the United States and Canada now protect prairie dogs.

Found in Africa and southern Asia, cheetahs are the fastest land animals in the world. With a small head, long legs, and a tail that helps them to keep their balance, cheetahs can run almost 60 miles (96 kilometers) per hour for short distances. Once hunted for their black-spotted fur, cheetahs are now protected. The major threat remaining for cheetahs is the loss of habitat, as more land is taken for farming and development.

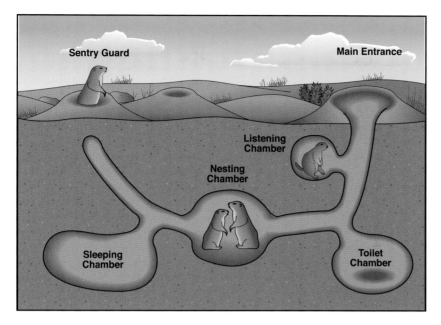

An illustration of a prairie dog "town." The underground tunnls are used to raise the young, hibernate in the winter, and to hide from predators.

Many species of kangaroos are found in Australia and on neighboring islands such as New Zealand. A number of the larger species are endangered. Some are killed for hides (skins) and some for food. Kangaroos are grazing animals, and many are shot by ranchers and farmers who want their cattle to graze without competition.

North American pronghorn antelopes are found in Mexico and the western United States. In the nineteenth century, about 40 million pronghorns lived on the prairies of North America. In the early twentieth century, much grassland was destroyed for farming, and many pronghorns were killed for food and hides. There are fewer than 10,000 pronghorns left and they live in the state of Arizona. Conservation laws have provided protected areas, but the pronghorn still faces the threat of extinction because of shrinking habitats.

A similar fate occurred with the American bison. Prior to the expansion of the United States westward, there were about 30-60 million bison. Their number was reduced to only about 1,000 in the late nineteenth century. The herds have since increased to nearly 30,000 on protected lands. The purity of the species is threatened as many bison are cross-bred with cattle for commercial purposes.

The New Look of Lawns

Using grass for lawns began in France and England in the late eighteenth century. It became popular in the United States, first as a way for people to tame wild-looking plots, and eventually as a status symbol—the green yard with no weeds.

Lawns have a high environmental cost. Gas lawnmowers pollute the air, thousands of gallons of fresh water are used up in watering, and tons of pesticides (insect poisons) are applied every year. Unfortunately, these chemicals kill wildlife and endanger the health of people and pets.

Many communities encourage the use of native trees, shrubs, flowers, and grasses to replace the typical lawn. Native plants do not require the same amount of water and fertilizer necessary to maintain lawn grasses. They also return the landscape to its original beauty and protect the environment.

Human Life

Grasslands have been home to people around the world for thousands of years. Of all the flowering plants, none is more important to human life than grasses. Grasslands dominate the agricultural regions of every continent except Antarctica.

Impact of the grassland on human life   Grasslands are essential to human life because of their role in food production and agriculture. Based on their usefulness to humans, there are six major types of grasses: grazing and forage grass, turfgrass, ornamental grass, cereal grass, sugarcane, and woody grass.

Food   More than half of the world's population relies on grasses for food. About thirty-five species of grasses are grown for human and animal use. Many grasses are harvested for hay, to feed livestock. Grasses are also cultivated for their seeds, which are processed and eaten. These grasses include the cereal grains: wheat, oats, barley, rye, corn, millet, sorghum, and rice. Wheat and corn are grown all over the world, with large crops coming from the United States, China, and Russia. Oats and rye are the chief domestic crops from Russia and Europe. Cereal, bread, and pasta are common foods made from these grains. Some wild grass species, such as wild rice, are also harvested. Sugarcane is a type of grass, but it is not a cereal grass. Its sap is concentrated into sugar. The sprouts and shoots of bamboo, which is a woody grass, also provide food for humans.

Grassland animals, such as the American bison (commonly known as the buffalo), have been used for food since the earliest times. When the horse, another grassland animal, became domesticated (tamed), hunters could follow wild herds of bison as they migrated (moved from one area to another). Overhunting is one reason the American bison has become an endangered species.

Economic factors   Many of the world's grasslands are used for farming and ranching and have great economic importance. Grazing and forage grasses are used for feeding cattle and other animals. Turfgrass is used to cover athletic fields, lawns, golf courses, and playgrounds. Turfgrasses include Kentucky blue grass, ryegrass, Bermuda grass, and buffalo grass. Ornamental (decorative) grasses, such as pampas grass and Chinese silvergrass, are used in parks and flower gardens. Sugarcane is used not only for sugar but also for making plastic and wallboard. Woody grasses, such as bamboo, can be used to make sturdy items including furniture and even houses. Fiber from the bark of the baobab tree is used for making rope and paper.

Grasslands are favorable sites for homes and cities because they tend to be relatively flat. Saskatoon in Saskatchewan, Canada, and Omaha, Nebraska, are both grassland cities.

Many grasslands offer mineral resources. Gold and diamonds are mined under the South African veld, while natural gas and petroleum are extracted from beneath the grasslands of Texas and Oklahoma.

Impact of human life on the grassland   At one time grasslands were the largest single biome in the world, covering more than 40 percent of Earth's land surface. None of these native grasslands remain untouched by humans.

When the railways brought large numbers of colonists to remote regions during the nineteenth century, grasslands were rapidly taken over for farms and towns. The Trans-Siberian Railroad brought people onto the steppes of Siberia. Similarly, railways crossed North America, moving large numbers of people all across Canada and the United States.

Since that time, many grasslands have become artificial, or human-made. Some of the grasslands were created in areas once occupied by forests that had been cut down. Almost all of the grasslands in Europe are artificial and are used for grazing animals and growing grain.

Use of plants and animals   Many native grassland plants are used for herbal medicines. The purple coneflower, for example, can be used to help heal wounds, and fleabane is used to repel insects. With the popularity of herbal remedies, many of these plants are being overharvested.

As more and more grassland is used for agriculture, native plants are destroyed leaving little food for wildlife. Trees are cut down and used for lumber or as fuel for cooking.

Grassland animals have also suffered from overuse. Mass slaughter and poaching of animals have nearly destroyed every major species of mammal and reptile and several species of birds in the South African veld.

After years of misuse, people have begun to restore prairies to their natural state. Here a refuge operations specialist at Squaw Creek National Wildlife Refuge works at clearing trees on a little over three acres of ground on the refuge. The refuge purchased the ground in the late 1980s and has been restoring the area to its original loess bluff prairie state. AP IMAGES.

Kruger National Park in South Africa is one of several protected areas in which some of these species survive.

Quality of the environment　During the 1930s, serious misuse of grassland resulted in the great "Dust Bowl" in Oklahoma and Texas. Farmers failed to use soil conservation methods and, while plowing for crops, destroyed the grasses that held the soil in place. After a long, dry period, the crops died and sweeping winds blew the dry topsoil away. The dust drifted thousands of miles and rose as high as 30 feet (9 meters). It was so thick in the air that people became sick from inhaling it. Crops could no longer be grown in the ruined land. Many people went hungry and thousands were forced to move.

Erosion (wearing away of soil) is still a problem in grasslands. African savannas are being stripped of their natural vegetation so they can be used for agriculture. In the Sahel region of North Africa, more and more grassland is becoming desert each year.

Some farmers use irrigation techniques to overcome dry grassland conditions. However, when water is pumped from underground sources, these water supplies can be completely used up.

In certain grasslands, measures are being taken to overcome the negative impact of human interference. Irrigation ditches linking fields to nearby aboveground sources bring water to crops, and farmers plow in ways that help prevent erosion. Native vegetation is protected from cattle so that it has a chance to regrow.

Native peoples Hunter-gatherers were living on the tropical grasslands of East Africa more than 40,000 years ago. About 10,000 years ago, these hunter-gatherers learned to cultivate the grasses and to domesticate animals, which allowed them to lead more settled lifestyles. Grasslands in some parts of the world continue to support tribes of people who live a traditional lifestyle.

Nomadic (wandering) hunters, herders, and shepherds once lived and roamed freely on temperate grasslands in North America. The Blackfoot, Teton Sioux, Cheyenne, and Comanche tribes followed the bison. Food, clothing, and shelter were all provided from the meat and skins of these animals. Some tribes, such as the Pawnee, Mandan, Hidatsa, and Arikara, were seminomadic and spent time planting and harvesting crops, especially corn, beans, and squash. By the nineteenth century European peoples began to move onto the world's grasslands, forcing the native peoples out of their traditional homes. Although some Native Americans still live on reservations (areas set aside for them to live), many have moved to large cities.

The Kazakh (also spelled Kazak) peoples live on the grasslands of Central Asia and parts of China. The Kazakh are nomads who raise sheep, cattle, goats, camels, and horses. Most still follow the traditional lifestyle, moving from one grassland to another when their animals need fresh pasture. Their food is primarily milk products and meat from their sheep. They live in portable, dome-shaped tents called yurts, which consist of a frame of poles covered with skins or wool felt. Many of their other needs are met with products from their animals: hides are made into clothing, horns are used for utensils, and horsehair is braided into rope.

African peoples still living on the grasslands include some tribes of Maasai found in the Great Rift Valley of southern Kenya and Tanzania. They are traditionally herders who move across the savannas with their cattle. Their cultural beliefs require them to live almost entirely off the

Learn More About Life on the Grassland

- Clark, Ann Nolan. *Secret of the Andes*. New York: Viking Press, 1952
- Friggens, Paul. *Gold and Grass, The Black Hills Story*. Boulder, CO: Pruett Publishing, 1983
- Hurwitz, Johanna. *A Llama in the Family*. New York: Morrow Books, 1994
- Low, Ann Marie. *Dust Bowl Diary*. Lincoln, NB: University of Nebraska Press, 1984
- Norton, Lisa Dale. *Hawk Flies Alone: Journey to the Heart of the Sandhills*. New York: Picador Press, 1996
- Warren, Andrea. *Pioneer Girl: Growing Up on the Prairie*. New York: Junior Books, 1998

Yurts are dome-shaped, portable shelters used by the nomadic Kazakh. IMAGE COPYRIGHT MARC VAN VUREN, 2007. USED UNDER LICENSE FROM SHUTTERSTOCK.COM.

meat, blood, and milk provided by their livestock. Most Maasai are permitted to eat grains and vegetables when food is scarce, but warriors are not. Maasai villages, called kraals, consist of mud-dung houses for four to eight families surrounded by a large circular thornbush fence. As less and less grazing land is available for the Maasai, many are forced to give up their traditional lifestyle and make their living trying to farm.

The Food Web

The transfer of energy from organism to organism forms a series called a food chain. All the possible feeding relationships that exist in a biome make up its food web. On the grasslands, as elsewhere, the food web consists of producers, consumers, and decomposers. An analysis of the food web shows how energy is transferred within a biome.

Green plants are the primary producers in the grassland. They produce organic (matter derived from living organisms) materials from inorganic chemicals and outside sources of energy, primarily the sun.

Animals are consumers. Primary consumers eat plants and include grazing animals, such as zebras, bison, and antelopes. In the savannas, where trees are more common, tall plant-eaters, such as giraffes and elephants, browse on the leaves of trees or bushes.

Plant-eaters become food for predators, which are secondary consumers such as coyotes and lions. Tertiary consumers eat both primary and secondary consumers. They include leopards and eagles. Humans are omnivores, which means they eat both plants and animals.

Decomposers eat the decaying matter from dead plants and animals and help return nutrients to the environment. For example, small underground insects called springtails help decomposition by breaking down dead plants. This allows other organisms, like bacteria and fungi, to reach the decaying matter and decompose it further.

Spotlight on Grasslands

The Llanos   The Llanos (YAH-nos) is a tropical savanna found in Venezuela and northeastern Colombia. The Llanos is bordered by the Andes Mountains in the north and west, the Amazon and Guaviare Rivers to the south, and the lower Orinoco River to the east. At its highest elevation, the Llanos is about 1,000 feet (305 meters) above sea level, although some areas may be as a low as 200 to 300 feet (61 to 91 meters).

Typical of tropical grasslands, the Llanos is flooded for part of the year and dry for the rest. During the dry season, which lasts from December through April, fires often sweep the land. When the rains begin in May, lakes, lagoons, and marshes form. The area stays wet until November, when the dry cycle begins again. Annual precipitation ranges from 30 to 98 inches (76 to 160 centimeters). The annual average temperature is approximately 80°F (27°C).

Swamp grasses, sedges, and carpet grasses grow throughout the region, which is also home to the Llanos palm, the scrub oak, cassias, and the araguaney tree.

During the wet season, the great anaconda inhabits the areas around the rivers. The anaconda is the largest snake in the Americas, measuring 30 feet (9 meters) long. A type of constrictor, the anaconda wraps itself around its prey and squeezes it. The Orinoco crocodile, one of the most threatened reptiles in the world, lives along the edge of the river.

Barbed Wire and the End of the Old West

Before the end of the nineteenth century, the American West was an open range where ranchers could graze cattle freely and cowboys could drive large herds to market. As farmers began to settle in the area they needed to protect their crops and land from wandering cattle. Cattle often trampled wooden fences in an effort to get to food or water so a stronger barrier was necessary.

In 1874, Joseph F. Glidden invented a machine to manufacture barbed wire. Barbed wire was made out of two strands of wire twisted together from which sharp barbs protruded. The barbs caused pain and kept cattle from damaging the fence.

As the open range was cut up by fences, "range wars" developed between farmers and ranchers. By 1888, the farmers had won, and cattle ranchers were forced to keep their cattle on their own property. They began to use railway cattle cars rather than long cattle drives over open land to transport them to market.

The Llanos
Location: Venezuela and northeastern Colombia in South America
Area: 220,000 square miles (570,000 square kilometers)
Classification: Savanna

Many species of birds live on the Llanos, including herons, storks, and ibises. The scarlet ibis, a large wading bird, has long legs and a long, slender, downward-curving bill used for searching in the mud for food. The jacana is a water bird whose long toes enable it to walk on the large floating leaves of water plants.

The capybara is the world's largest rodent, weighing about 100 pounds (45 kilograms). It eats grasses and aquatic herbs, which it finds both on land and in the water. The capybara is a favorite meal for both the anaconda and the Orinoco crocodile. The only wild animals with hooves that live on the Llanos are the white-tailed deer and the red brocket, a small deer with short, unbranched antlers.

The Llanos is home to the giant anteater and giant armadillo. A toothless animal that eats insects, but is especially fond of ants and termites, the giant anteater is 6 to 8 feet (1.8 to 2.4 meters) in length and weighs about 65 to 140 pounds (29 to 64 kilograms). The giant armadillo can weigh up to 100 pounds (45 kilograms) and is protected by a type of armor that covers its body from head to tail. In some species, the armor is in segments and is flexible. Unlike certain members of the armadillo family, the giant armadillo cannot roll into a ball to protect itself from predators. Instead, the giant armadillo quickly digs itself into the ground using the long claws on its front legs. Very shy and almost blind, it depends on its senses of smell and hearing. With its large claws, it rips apart insect nests in search of food but also eats roots and worms.

Serengeti National Park   The name Serengeti is from a Maasai word meaning "endless plains." Serengeti National Park and Wildlife Refuge is located in north central Tanzania on a high plateau between the Ngorongoro highlands and the Kenya/Tanzania border. Stretching between Lake Victoria and Lake Eyasi, the park is best known for the millions of wild animals that live there.

The park contains a vast open grassland in the southern region, a large acacia savanna in the center, and wooded grasslands in the north. Within the Serengeti are kopjes (rocky hills) with their own unique biomes. Rivers, lakes, and swamps are scattered throughout the park, providing habitats for a variety of reptiles and birds.

Located in the tropics, the Serengeti is dry and warm. Yearly temperatures range from 59° to 77°F (15° to 25°C), and the coolest weather is from June to October. The major rainy season lasts from March to May, but shorter periods of rain occur in October and November. About 47

Serengeti National Park
Location: North central Tanzania in Africa
Area: 5,700 square miles (14,763 square kilometers)
Classification: Savanna

inches (120 centimeters) of rain fall near Lake Victoria and about 20 inches (51 centimeters) on the plains.

Elevations in the park range from 3,000 to 6,000 feet (920 to 1,850 meters).

Not many amphibians or reptiles inhabit grasslands. However, in 1992, a new species of tree frog was discovered in the Serengeti's rocky hills during the rainy season. This species is named *Hyperolius orkarkarri.*

More than 500 species of birds have been seen in the park. Common birds include starlings, ring-necked doves, and the barbet, a brightly colored bird with a large, strong bill with bristles at its base.

About thirty-five species of herbivores, such as wildebeests, zebras, giraffes, gazelles, topi, and konga, live here. Elephants began moving onto the Serengeti in the 1960s when human populations in the surrounding area increased, forcing them from their former homes. Carnivores include lions, cheetahs, hyenas, and leopards.

Land animals in the Serengeti migrate annually. The migration follows the pattern of the seasons and is led by the millions of wildebeests in the park. During the wet season, which lasts from December to May, herds graze on the southeastern plains. As the season progresses, they move west into woodland savanna and then north into the grasslands as their food and water supplies become scarce. When the dry season ends in November, the herds return to the rain-drenched southeastern plains, and the cycle begins again.

The Maasai were the first people to inhabit the Serengeti, arriving with their herds at the end of the nineteenth century. In the early twentieth century, Europeans and Americans arrived and began the wholesale killing of wild animals. The lion population was almost wiped out, but game reserves were established in the 1920s, and the entire area was made into a park in 1951. Although the Serengeti is one of the largest wildlife sanctuaries in the world, poachers (people who hunt or fish illegally) still pose a major threat. Elephants are killed for their ivory tusks and the black rhinoceros, which is almost extinct, is killed for its horn.

Wind Cave National Park   Wind Cave National Park is located in South Dakota and borders the ponderosa pine forests of the Black Hills. It is home to a mixed-grass prairie that contains both tall grasses from the eastern prairies and short grasses from the high plains near the Rocky

Wind Cave National Park
Location: South Dakota
Area: 28,292 acres
(11,449 hectares)
Classification: Mixed-grass prairie

Mountains. The park was established in 1903 to preserve its grasslands and its many limestone caverns.

The climate in the park is typical of a temperate grassland—warm in summer and cold in winter. The air that comes off the Rocky Mountains makes the park somewhat warmer and drier than the surrounding areas. In winter, temperatures range between 22° and 50°F (-5° and 10°C). Summer temperatures average 85°F (20°C). Some moisture comes from snow, which can accumulate as much as 30 inches (76 centimeters) in the winter.

Most of the vegetation in the prairie consists of grasses. When rainfall is more plentiful, the tall grasses dominate because they grow best in a moist environment. When there is less rain, the short grasses take over. Native tallgrass species include spikebent, redtop, big bluestem, and prairie sandreed. Native mixed-height grasses include slender wheatgrass, little bluestem, and Junegrass. Red three-awn, buffalo grass, and stink-grass are native short grasses. Other native vegetation includes forbs, such as small soapweed (a yucca plant), prairie clover, and Indian hemp dogbane.

Common amphibians that make this park their home include the blotched tiger salamander, the plains spadefoot toad, the upland chorus frog, and the Great Plains frog, which is easily identified by its many bumps. Reptiles found along the streams include the common snapping turtle, the wandering garter snake, and the prairie rattlesnake.

Hundreds of species of birds visit or make the park a permanent home. Permanent residents range from common finches to hawks, golden eagles, and prairie falcons.

Mammals frequently seen in the park are coyotes, prairie dogs, chipmunks, and elks. Bison, elk, and pronghorns were reintroduced to the park in 1913. There are sightings of coyote, mountain lions, and whitetail deer, but black bears, grizzly bears, and grey wolves are no longer found.

Manas National Park   Manas National Park, located in the foothills of the Himalaya Mountains in India, is part forest and part grassland. It is home to a great variety of vegetation and wildlife. To the north is the country of Bhutan, to the south is North Kamrup, and forest preserves are to its east and west. The park includes part of Manas Reserve Forest and all of North Kamrup Reserve Forest. The Manas River runs through it. Manas was declared a wildlife sanctuary in 1928 and upgraded to a

Manas National Park
Location: Western Assam State in eastern India
Area: 125,000 acres (50,000 hectares)
Classification: Savanna

National Park in 1990. In 1992, it was recognized as a World Heritage Site in Danger because of heavy poaching and political unrest.

Much of the park is low-lying and flat, ranging in altitude from 100 to 361 feet (61 to 110 meters) above sea level. Summer is warm, lasting from April through June with a maximum temperature of 99°F (37°C). October through March is the chilly season with temperatures falling no lower than 51°F (11°C). The monsoon season lasts from May to September and is fairly warm. Annual rainfall, most of which falls during this season, is 131 inches (333 centimeters).

Many species of grasses and a variety of trees and shrubs are found in the park. In the northern forest region, evergreen and deciduous trees (trees that lose their leaves) grow. Tall, dense grasses used to manufacture paper are an important resource at these lower altitudes.

A few amphibian species and about 30 species of reptiles can be found in the park sanctuary. Reptiles include the vine snake, flying snake, Assam trinket snake, monitor lizard, and roofed turtle.

Hundreds of species of birds live in the park, including the great pied hornbill, the pied harrier, and the spot-billed pelican.

Elephants, hog deer, and tigers are among the more than fifty species of mammals in the sanctuary. The pigmy hog and the rare golden langur, a long-tailed monkey with bushy eyebrows and a chin tuft, rely on the park for their survival.

Eurasian steppes   One of the world's largest temperate grasslands is the Eurasian steppe, which extends about 5,000 square miles (8,000 square kilometers) across Hungary, Ukraine, Central Asia, and Manchuria. The steppes are bisected (divided) by the Altai Mountains into the western steppe and the eastern steppe. The western steppe stretches from the mouth of the Danube River along the north shore of the Black Sea and across the lower Volga River. The eastern steppe continues to the Greater Khingan Mountain Range. The entire terrain is criss-crossed by rivers and streams.

In the summer, temperatures average 73°F (23°C) June through August. The winter is cold and the area is covered with as much as 4 inches (10 centimeters) of snow. Average winter temperatures from November to March are below freezing; temperatures average 29°F (-2°C). Annual rainfall for the steppes averages between 10 and 20 inches (25 and 51 centimeters).

Eurasian Steppes
Location: Hungary, Ukraine, Central Asia, Manchuria
Area: 5,000 square miles (8,000 square kilometers)
Classification: Steppe

The steppes contain some of the most fertile land in the world. Primarily made up of black earth (chernozem), steppe soil is excellent for crops. Most of Russia's grain, for example, is produced on the steppes.

Native steppe vegetation consists primarily of turf grasses, such as bluegrass, bunchgrass, feather grass, and fescue, as well as mosses and lichens. In the north where there is more moisture, wild tulips, irises, daisies, and sages grow. Fewer flowers grow in the drier southern region.

Common reptiles found on the Eurasian steppe include steppe vipers and whip snakes. Birds include larks, bustards, and kestrels. An unusual bird that comes to the steppes to mate is the demoiselle crane, famous for its mating dance. About 3 feet (0.9 meter) in height, this blue-gray bird is characterized by tufts of feathers on the sides of its head. Birds of prey include hawks, falcons, and eagles.

Burrowing animals, like marmots, steppe lemmings, and mole rats, do well on the open steppes. The spotted suslik, a ground squirrel, lives in underground colonies. Other mammals that live on the steppes include skunks, foxes, wolves, antelopes, muskrats, raccoons, beavers, and silver foxes. Saigas, at one time on the verge of extinction, have recovered and are now found almost everywhere throughout the regions.

For thousands of years people have lived on the steppes, beginning with hunter-gatherers in prehistoric times. Eventually the lifestyle changed to farming and, about 2000 BC, when the horse was domesticated, people began herding animals and moving from pasture to pasture. Conflict between tribes was common until the sixteenth century when the Russians conquered and colonized the area. A few small groups of nomads still make their home on the steppes.

Parts of the steppes are being preserved as national parks and wildlife refuges. For example, the Askaniya-Nova in the Ukraine works to protect endangered species. More than forty different mammals, including the onager (wild ass) and Przewalski's horse, have been introduced to the park.

Konza Prairie Preserve The Konza Prairie Preserve in Kansas is a tallgrass prairie, a rich environment that plays host to over 100 ecological research projects at any given time. In addition to grasslands, the Konza area contains streams and a deciduous forest. Wide expanses of tallgrass are broken up by natural depressions in the ground, such as prairie potholes. After the rain, these areas fill up and provide watering holes for animals.

The Konza climate is temperate, with warm, moist summers and cool, dry winters. Summer temperatures range from 80° to 100°F (27° to 38°C), while winter temperatures can be as low as 10°F (-12°C). Annual average rainfall is 32 inches (81 centimeters).

The most common grass on the Konza is big bluestem. Other grasses include Indian grass, switchgrass, and little bluestem. Forbs such as asters and sunflowers are abundant, while sedges are less common. Many prairie plants once used by Native Americans for food and medicine still grow in the Konza, including wild nodding onion, wild plum, prairie turnip, and prairie parsley.

Of the many different insect species found on the Konza, June beetles, dung beetles, lappet moths, butterflies, and grasshoppers are perhaps the most common.

Few amphibians live on this prairie since there are no natural ponds. Depressions that fill with water and artificial ponds originally built to provide water for livestock do provide homes for some species of frogs, toads, and salamanders. The bullfrog is the most common.

Reptiles roaming the Konza include the western box turtle and the collared lizard. The most common reptile is the Great Plains skink, a type of lizard that can detach its tail when attacked and leave it wriggling on the ground to confuse the predator. A new tail grows in quickly.

The open prairie is home to birds that nest on the ground, such as meadowlarks and mourning doves. Cowbirds have an unusual adaptation to life on the open prairie: they lay their eggs in the nests of other birds, who raise the cowbird babies with their own.

Mammals living in the preserve include a large bison herd and small groups of deer. Many rodents live here, with deer mice being the most plentiful. Voles and shrews, some of the smallest mammals in the world, also make the preserve their home.

The Konza Preserve is as a conservation partnership. Most of it is owned by The Nature Conservancy (an organization that establishes private nature sanctuaries in order to preserve plants, animals, and natural communities) and serves as an outdoor research laboratory run by the Division of Biology at Kansas State University. The area was chosen by the National Science Foundation as a long-term ecological research site.

Konza Prairie Preserve
Location: Riley and Geary Counties in Kansas
Area: 8,616 acres (3,487 hectares)
Classification: Tallgrass prairie

For More Information

BOOKS

Allaby, Michael. *Biomes of the Earth: Grasslands.* New York: Chelsea House, 2006.

Grzimek, Bernhard. *Grizmek's Animal Encyclopedia*, 2nd edition. Volume 7. *Reptiles,* edited by Michael Hutchins, James B. Murphy, and Neil Schlager. Farmington Hills, MI: Gale Group, 2003.

Hancock, Paul L., and Brian J. Skinner, eds. *The Oxford Companion to the Earth.* New York: Oxford University Press, 2000.

Humphreys, L.R. *The Evolving Science of Grassland Improvement.* Cambridge, UK: Cambridge University Press, 2007.

Luhr, James F., ed. *Earth.* New York: Dorling Kindersley in association with The Smithsonian Institute, 2003.

Moul, Francis. *The National Grasslands: A Guide to America's Undiscovered Treasures.* Lincoln: University of Nebraska Press, 2006.

PERIODICALS

Cunningham, A. "Going Native: Diverse Grassland Plants Edge Out Crops as Biofuel." *Science News.* 170. 24 December 9, 2006: 372.

De Silva, José María Cardoso, and John M. Bates. "Biogeographic Patterns and Conservation in the South American Cerrado: A tropical Savanna Hotspot." *BioScience* 52.3 March 2002: p225.

Donaldson, Mac. "Corridors for Migration." *Endangered Species Bulletin.* 28. 3 May-June 2003: 26.

Flicker, John. "Audubon view: Grassland Protection." *Audubon.* 107. 3 May-June 2005: 6.

Hoekman, Steven T, I.J. Ball and Thomas F. Fondell. "Grassland Birds Orient Nests Relative to Nearby Vegetation." *Wilson Bulletin.* 114. 4 December 2002: 450.

Kloor, Keith. "Fire (in the Sky): in Less than an Hour, Flames had Reduced Nearly 8,000 Acres of Grasslands to Smoldering Stubble and Ash." *Audubon.* 105. 3 September 2003: 74.

Springer, Craig. "Leading-edge Science for Imperiled, Bonytail." *Endangered Species Bulletin.* 27. 2 March-June 2002: 27.

ORGANIZATIONS

Environmental Protection Agency, 401 M Street, SW, Washington, DC 20460, Phone: 202-260-2090; Internet: http://www.epa.gov (accessed August 17, 2007).

Friends of the Earth, 1717 Massachusetts Ave. NW, 300, Washington, DC 20036-2002, Phone: 877-843-8687; Fax: 202-783-0444; Internet: http://www.foe.org.

Greenpeace USA, 702 H Street NW, Washington, DC 20001, Phone: 202-462-1177; Internet: http://www.greenpeace.org.

Nature Conservancy, Worldwide Office, 4245 North Fairfax Drive, Arlington, VA 22203-1606, Phone: 800-628-6860; Internet: http://www.nature.org.

Sierra Club, 85 2nd Street, 2nd Fl., San Francisco, CA 94105, Phone: 415-977-5500; Fax: 415-977-5799, Internet: http://www.sierraclub.org.

World Meteorological Organization, 7bis, avenue de la Paix Case postale No. 23000, CH-1211 Geneva 2, Switzerland, Phone: 41(0) 22 7308111; Fax: 41(0) 22 7308181, Internet: http://www.wmo.ch.

World Wildlife Fund, 1250 24th Street NW, Washington, DC 20090-7180, Phone: 202-293-4800; Internet: http://www.wwf.org.

WEB SITES

Blue Planet Biomes: http:http://www.blueplanetbiomes.org (accessed September 14, 2007).

Envirolink: http://www.envirolink.org (accessed September 14, 2007).

"The Grassland Biome." University of California Museum of Paleontology. http://www.ucmp.berkeley.edu/exhibits/biomes/grasslands.php (accessed September 14, 2007).

National Geographic Magazine: http://www.nationalgeographic.com (accessed September 14, 2007).

National Park Service: http://www.nps.gov (accessed September 14, 2007).

"Wind Cave National Park." The National Park Service. http://www.nps.gov/wica/ (accessed September 14, 2007).

Lake and Pond

Lakes and ponds are inland bodies of water. Ponds tend to be shallow and small, and most do not have names. Lakes vary greatly in size and most do have names. Lake Superior, for example, which lies between Canada and the United States, has the greatest surface area of any freshwater lake in the world—31,800 square miles (82,362 square kilometers). Lake Baikal in southern Siberia is the deepest at 1 mile (1.6 kilometers). Baikal holds the most water even though its surface area is less than half that of Lake Superior. Ponds and lakes differ in overall water temperature. Ponds have a uniform temperature throughout, but a lake has two distinct layers: an upper layer affected by air temperature, and deeper water that may be either warmer or colder than the upper layer, depending upon the season.

Less than 1 percent of all the water on Earth is held in lakes and ponds. Even so, the amount is considerable. In North America alone, there are at least 1.5 million ponds, totaling as much as 2 million acres (808,000 hectares) of water. Both lakes and ponds tend to be more numerous in the northern hemisphere and in mountain regions. There are relatively few in South America, for instance, but the northern areas of Canada are a large network of lakes and ponds.

How Lakes and Ponds Develop

All lakes and ponds have a life cycle that begins when they are formed and ends when they are filled in with plant life.

Formation   Both lakes and ponds usually form when water collects in undrained depressions, or basins, in the ground and any outlet, such as a stream, does not drain them completely. The source of the water may be precipitation (rain, sleet, and snow), a river, a stream, a spring, or a melting glacier. In any case, there must be enough water to keep the depression

WORDS TO KNOW

Epilimnion: The layer of cold or warm water closest to the surface of a large lake.

Euphotic zone: The zone in a lake where sunlight can reach.

Eutrophication: Loss of oxygen in a lake or pond because increased plant growth has blocked sunlight.

Herbivore: An animal that eats only plant matter.

Hypolimnion: The layer of cold or warm water closest to the bottom of a large lake.

Kettle: A large pit created by a glacier that fills with water and becomes a pond or lake.

Parasite: An organism that depends upon another organism for its food or other needs.

Profundal zone: The zone in a lake where no more than 1 percent of sunlight can penetrate.

Sediments: Small, solid particles of rock, minerals, or decaying matter carried by wind or water.

Seiche: A wave that forms during an earthquake or when a persistant wind pushes the water toward the downward end of a lake.

Soda lake: A lake that contains more than 0.1 ounce of soda per quart (3 grams per liter) of water.

filled. In very dry regions, ponds or lakes may form during a rainy season and then disappear when the dry season returns.

Depressions that become ponds and lakes may be created by natural forces, animals, people, and even the wind. Most are made by glacial, volcanic, tectonic (crustal plate movement), or riverine action, and by sinkholes and barriers.

Glacial action   During the Ice Ages, starting 2 million years ago, glaciers (gigantic slow-moving rivers of ice) gouged depressions in the land. Many of these depressions filled with water, creating lakes and ponds. Examples include lakes in Canada, the northern United States, Finland, and parts of Sweden. Glaciers on the tops of mountains, such as in Glacier National Park in Montana, still produce lakes by means of the same process, and most mountain lakes originated in this way. Berg Lake in Canada gets its name from the icebergs that break off a glacier sitting along its shoreline. Glacial basins tend to be shallow and rimmed by rocky shorelines.

When glaciers melt, they sometimes leave pits, called kettles, which fill with the meltwater and become ponds, lakes, or wetlands. Prairie regions in North America have many of these kettles, also called prairie potholes.

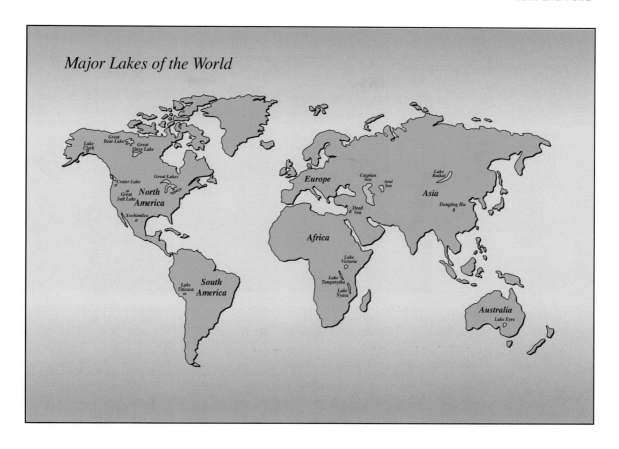

Major Lakes of the World

Volcanic action   The craters of extinct or inactive volcanoes often contain lakes. One example is Crater Lake in Oregon. The walls of laval rock surrounding this lake rise to 1,932 feet (589 meters) above its surface. Lakes in craters often contain islands. Some of these islands are cones of ash and lava created when the volcano went through another active period. The largest crater lake in the world is Lake Toba in Sumatra, although it is undergoing changes due to earthquake activity.

Tectonic action   Tectonic action refers to movement of Earth's crust during earthquakes. Earth's crust is always in motion causing some pieces to push against one another and others to pull apart. During earthquakes, great cracks may form in the ground. When water fills these cracks, it forms lakes. Examples of lakes created by tectonic action include Lake Tahoe in California and Lake Baikal in Siberia.

Wizard Island is a small volcanic cone in Crater Lake, which is in a crater of an extinct volcano in southern Oregon. IMAGE COPYRIGHT BRYAN BRAZIL, 2007. USED UNDER LICENSE FROM SHUTTERSTOCK.COM.

Riverine action   During floods, rivers may overflow their banks, creating shallow floodplain lakes. The Amazon River in Brazil, for example, frequently becomes flooded and thousands of lakes form along its great length. Winding rivers often create oxbow lakes during floods. The river jumps its banks and changes course. The abandoned section, which is curved, or bow shaped, becomes a lake.

Sinkholes   Sinkholes are formed when the underlying rock is limestone and a source of water, such as an underground river, dissolves the limestone. The ground overhead collapses, leaving a pit. If the pit fills with water, a lake or pond is formed. Sinkhole lakes may be 50 feet (15 meters) or more deep. Several sinkhole pools, called Silver Springs, are located in northern Florida.

Barriers   Barriers of sand, gravel, mud, rock, lava, and glacial debris can dam up a river or stream, and the area behind the dam fills with water. This type of lake is often short lived because the movement and pressure of the water soon cuts through the debris and the lake is drained.

Succession   Lakes and ponds are not permanent, even though some lakes may exist for thousands of years. A volcano or tectonic action, for example, may change the structure of the underlying rock, causing the water to drain away. Streams or other sources that feed the lake may change course or dry up, or the climate may change, yielding little precipitation. Most lakes and

The Killer Lakes

In the Cameroon highlands of West Africa is a series of beautiful but deadly lakes that are responsible for killing nearly 2,000 people during the 1980s. Rolling hills covered in tall grasses and lush vegetation are characteristics of this region, and the soil is so rich that many farming communities of the Bantu and Fulani tribes have located there. Legends of the Bantu and Fulani tell of strange floods that destroyed villages and of spirit women who live in the lakes and pull people to their deaths. There is some truth to these legends.

In 1984 the police in a nearby village were told about people dying on the road near Lake Monoun. When they went to investigate, they saw a smokelike cloud drifting outward from the lake that left seventeen people dead in its wake. Lake Monoun and its sister lakes are pools that formed in the craters of inactive volcanoes. It was determined that a small earthquake had shaken up the lower depths of the lake where carbon dioxide gas was suspended like bubbles in a can of soda. All at once these deep waters were forced upward to where the pressure was less, and the gas was released into the air. In high concentrations, carbon dioxide is deadly, as it was in this case when it suffocated everyone in its path.

Before the mystery was solved and the information was made available, similar clouds had erupted from other lakes in the region, taking many lives.

ponds begin to transform as soon as they are created because they gradually fill with sediments (particles of soil and other matter) and dead plant and animal matter. This ecological process is called succession.

Succession usually follows the same pattern. As sediments fill the bottom of the pond or lake, shore plants, such as reeds and grasses, can gain a foothold and grow. Gradually, they begin to move in toward the center of the lake. The increased waste from dead stems and leaves makes the water thick, shallow, and slow moving. Eventually, shrubs and then trees, such as willows, begin to grow. The open water continues to shrink until the area becomes a wetland. As succession continues, the wetland disappears and is replaced by dry ground.

The speed with which succession takes place varies, depending upon the kinds of sediments and the death rate of plants in the region. A small pond can disappear in fewer than 100 years; a lake takes much longer, perhaps several thousand years.

Kinds of Lakes and Ponds

Lakes and ponds can be classified in many ways. One of the most common ways is by chemical composition. They may contain fresh water, salt water, or soda water.

Water Down Under

Groundwater is what its name implies—water beneath the ground. Over time, water from ponds, lakes, and rain or snow trickles down through the earth and collects between layers of rock. When someone digs a well, the water that fills the well is groundwater.

A spring occurs when groundwater breaks through to the surface. Springs may feed lakes, ponds, and natural wetlands. The world's largest reported spring is Ras-el-Ain in Syria, with an average yield of 10,200 gallons (38,700 liters) per second. The largest spring in the United States is Silver Springs in Florida, which averages 6,100 gallons (23,000 liters) per second.

Freshwater lakes and ponds   The water in freshwater lakes is relatively pure. It contains many dissolved minerals and salts at very low concentrations. Freshwater lakes tend to be located in temperate or cold regions, and they support much plant and animal life.

Saltwater lakes and ponds   The salt found in salt lakes and ponds may be sodium chloride, which is ordinary table salt, or it may include a combination of other types, including magnesium salt. The Great Salt Lake in Utah contains sodium chloride, whereas the Dead Sea between Israel and Jordan contains a combination of salts.

These salts usually enter the water as it dissolves the surrounding rock, or they are carried in by streams. Salt lakes occur primarily in dry climates where evaporation is sped up and the concentration of salt is able to build.

Salt lakes contain at least 0.1 ounce (3 grams) of salt per quart (liter) of water. The minimum amount produces water less salty than seawater, but the concentration in some lakes makes them much saltier than the ocean. The Dead Sea, which has the highest salt content of any lake on Earth, is nine times saltier than ocean water. When these bodies of water dry up, they leave behind a crust of salt, sometimes referred to as white alkali.

Soda lakes and ponds   Soda (alkali) lakes and ponds contain minerals that are usually produced in hot, volcanic springs. The primary mineral is sodium bicarbonate, which is similar to baking soda. Lake Natron in Tanzania is an example of a soda lake. Like salt lakes, soda lakes usually occur in hot climates. The water in Lake Natron evaporates quickly under midday temperatures as high as 140°F (60°C), but it is fed by geysers (volcanic springs) that help maintain its size.

To be classified as a soda lake, a lake must contain more than 0.1 ounce (3 grams) of soda per quart (liter) of water. The soda in Lake Natron is so concentrated that in places it is thick and hard enough to walk on. Natron's water kills almost all plant and animal life and can burn human skin seriously enough to require surgery.

The Water Column

The water column refers to the water in a lake or pond, exclusive of its bed or shoreline.

Composition   The water in a pond or lake may be fresh, salty, or alkaline; it may be clear or cloudy; it may be polluted or clean. Its characteristics are determined by where it comes from and the nature of its bed or basin. For example, if a river flowing into a lake carries a large quantity of sediment with it, the water in the lake will be muddy.

Zones   Different parts of a lake or pond may have different features and support different kinds of plants and animals. These different parts are called zones. Zones may be determined by temperature, vegetation, or light penetration.

Zones determined by temperature   Ponds and small lakes often have a uniform temperature throughout because they are relatively shallow. The water in large lakes may form layers based on water temperature. This difference in

Pond Inspiration

Many books have been written about ponds, but few compare to *Walden* by Henry David Thoreau (1817–1862). Thoreau, a famous American writer, wrote his book after living two years beside Walden Pond, a small body of water near Boston, Massachusetts.

Born in 1817, Thoreau was an artist whose own life became the material for his book. In a society rapidly growing more urban and industrial, he believed in the wholeness of nature and in a life of principle. In 1845 he sought solitude and simple living beside Walden Pond, where he could think, write, and discover "the great facts of his existence." He built a small cabin and did some farming. The journals he kept became the basis for his book, which he published in 1854.

The importance of *Walden* and Thoreau's other works were not appreciated during his lifetime. These records of what was basically a spiritual journey are now among the world's finest writings.

Fishermen rowing their boat in Lake Baikal in Siberia, Russia. Lake Bailkal is the deepest lake in the world. IMAGE COPYRIGHT VOVA POMORTZEFF, 2007. USED UNDER LICENSE FROM SHUTTERSTOCK.COM.

Geysers

Deep within Earth is a core of molten rock that, in places, is close enough to the surface to heat groundwater and turn it into steam. When this steam builds up, it forces its way to the surface in a gush of water and vapor called a geyser (GY-zuhr).

Geysers and hot springs usually occur in regions where volcanoes were or are active, and some, such as Old Faithful in Yellowstone National Park, erupt with clocklike regularity. Pools of hot mineral-rich mud may form in the same region. These naturally hot springs and pools were used by humans as early as 190 BC for their healing properties.

Since 1900, scientists have experimented with tapping the heat within Earth for use as a power source. Called geothermal (jee-oh-THUR-muhl) energy, it can be used to drive engines and heat water. In Iceland, many homes and greenhouses are heated using geothermal energy.

temperature prevents the waters from mixing well. In temperate climates (those with warm summers and cold winters) these layers occur seasonally.

In the summer the heat of the sun and the warmth of the air create an upper layer of warm, circulating water called the epilimnion (eh-pih-LIHM-nee-uhn). The colder water, which is heavier and noncirculating, sinks to the bottom where it forms a layer called the hypolimnion (HIGH-poh-lihm-nee-uhn). The zone in between is called the thermocline.

In the autumn, heat in the epilimnion is lost as cooler weather moves in, and the lake achieves about the same temperature throughout. In winter, the epilimnion is exposed to the cold air temperatures and may freeze, forming a layer of ice, while the hypolimnion remains comparatively warm. The layers do not mix again until the spring when the ice thaws.

Zones determined by vegetation Along the edges of a pond or lake, areas of vegetation run somewhat parallel (in the same direction) to the shoreline. Here the soil is marshy and wet. The area where shallow water begins to appear around the roots of plants marks the beginning of the littoral zone. This zone is the area near the shore where plants are rooted at the bottom and light penetrates down to the sediment. This zone supports a large variety of animal life. The width of the littoral zone may vary from a few feet to a few miles.

The limnetic zone is the deeper, central region characterized by open water and no vegetation.

Zones determined by light penetration Deep lakes may be divided into zones based on light penetration. The upper, or euphotic (yoo-FOH-tik) zone is exposed to sunlight and supports the most life. In clear lakes this zone may reach as deep as 165 feet (50 meters). In muddy lakes, the euphotic zones may only be 20 inches (51 centimeters) deep.

The lower, or profundal (proh-FUN-duhl) zone receives no more than 1 percent of sunlight. No plants grow here, although animals frequent this zone.

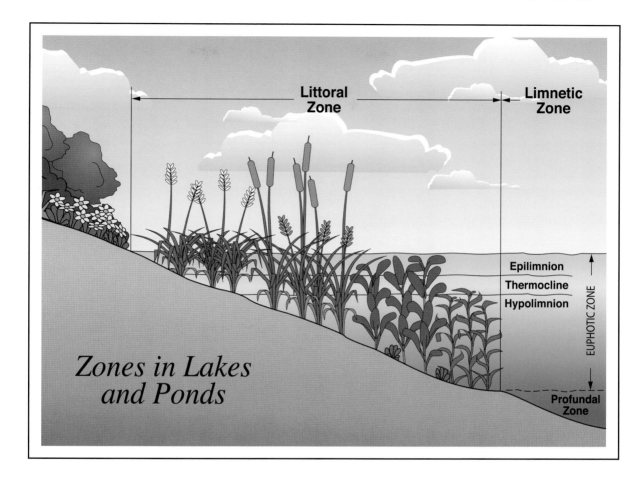

Littoral Zone

Limnetic Zone

Epilimnion
Thermocline
Hypolimnion

EUPHOTIC ZONE

Profundal Zone

Zones in Lakes and Ponds

Circulation   Water in a lake is usually in motion, forming both waves and currents.

Waves   Waves are rhythmic rising and falling movements in the water. Although waves make the water appear as if it is moving forward, forward movement is actually very small. Most surface waves are caused by wind. Their size is due to the speed of the wind, the length of time the wind has been blowing, and the distance over which it travels. As these influences grow stronger, the waves grow larger.

A seiche (SAYSH) is a wave that forms during an earthquake or when a persistent wind forces the water toward the downwind end of the lake. When the wind ceases, the water flows back and forth from one end of the lake to the other in a rocking motion.

Waves collapse on the shoreline because the water at the bottom of them is slowed by friction as it rolls along the shore. The top water then outruns the bottom and topples over. IMAGE COPYRIGHT ALEXANDER KOLOMIETZ, 2007. USED UNDER LICENSE FROM SHUTTERSTOCK.COM.

Currents   Currents are the steady flow of water in a certain direction. Surface currents are caused by persistent winds, the position of landmasses, and water temperature variations.

Winds tend to follow a regular pattern. They generally occur in the same place and blow in the same direction, and the movement of water follows this pattern. When a current meets a landmass, such as an island, it is deflected and forced to flow in a new direction.

Both horizontal and vertical currents below the surface occur when seasonal temperature changes cause layers of water (the epilimnion and hypolimnion) to form. As the layers take shape or break up, the water turns over (warmer water rises and the cooler water sinks). Most freshwater lakes turn over at least once a year and some twice. In salt and soda lakes, the heating and cooling of the upper layer is not strong enough to cause the water to turn over.

Effect of the water column on climate and atmosphere   The climate surrounding a pond or lake depends upon where the lake or pond is located. If it is located in a desert country, such as Iran, the climate will be hot and dry. If it sits on top of a mountain in the northern hemisphere, the climate will be cold. Ponds in temperate climates are often affected by seasonal changes. A small pond may completely dry up in summer, or

THE WORLD'S LARGEST LAKES

Lake	Location	Surface Area		Depth	
		Square Miles	Square Kilometers	Feet	Meters
Caspian Sea*	Azerbaijan, Kazakhstan, Turkmenistan, Iran, Russia	144,000	373,000	3,264	995
Lake Superior	Canada, United States	31,820	82,732	1,333	406
Lake Victoria	Kenya, Tanzania, Uganda	26,828	69,484	270	82
Lake Huron	Canada, United States	23,000	59,600	750	229
Lake Michigan	United States	22,300	57,757	923	281
Aral Sea*	Kazakhstan, Uzbekistan	15,500	40,100	177	54
Lake Tanganyika	Burundi, Zambia, Tanzania, Zaire	12,700	32,893	4,708	435
Lake Baikal	Siberia	12,162	31,494	5,315	620
Great Bear Lake	Canada	12,096	31,328	1,299	396
Lake Nyos	Tanzania, Mozambique, Malawi, Cameroon	11,100	28,749	2,300	701

*Saltwater lakes

freeze to the bottom in winter. Severe spring floods may destroy shoreline plants including the homes and feeding grounds of many animals.

The presence of a large lake can itself create some climatic differences. A large body of water absorbs and retains heat from the sun. In the winter, this stored heat is released into the atmosphere helping to keep winter temperatures around the lake warmer. In summer, when the water temperature is cooler than the air temperature, winds off the lake help cool the nearby region. This is known as the lake-effect. The climate of Chicago, which is located on the shores of Lake Michigan, benefits in this way.

The water in lakes and ponds is partly responsible for the precipitation (rain, sleet, or snow) that falls on land. The water evaporates in the heat of the sun, forms clouds, and falls elsewhere. Large lakes can even influence storms. In winter, for example, as air flows over a lake, it picks up moisture. When the warmer, moister air meets the colder, drier air over land, it produces lake-effect snow. Towns along the eastern shore of Lake Ontario regularly receive 118-157 inches (300-400 centimeters) of snow a year for this reason.

Bodies of water help regulate the levels of different gases in the atmosphere, such as oxygen and carbon dioxide. Too much carbon dioxide in the atmosphere contributes to warmer global temperatures. The presence of ponds and lakes helps moderate such undesirable changes by absorbing some carbon dioxide.

Geography of Lakes and Ponds

The geography of lakes and ponds involves the process of erosion (wearing away) and deposition (dep-oh-ZIH-shun; setting down), which helps determine the different types of shoreline surfaces and landforms.

Erosion and deposition   As waves slap against a shoreline they compress (squeeze) the air trapped in the cracks in rocks. As the waves retreat, the air pressure within the rocks is suddenly released. This process of pressure and release widens the cracks and weakens the rocks, causing them to eventually break apart. Waves created by storms in a large lake can be high and forceful. In places where wave action is strong, the water stirs up particles of rock and sand from the lake bed throwing them against the shoreline. As a result, particles in the water produce a cutting action, eroding the shoreline even further.

Some of the chunks and particles eroded from a shoreline may sink to the lake bed. Others may be carried by currents farther along the shore and deposited where there is shelter from the wind and the wave action is less severe. Erosion and deposition can change the geography of a shoreline over time.

Shoreline surfaces   The general shape of the shoreline is usually determined by the shape of the surrounding landscape. If the lake or pond is on a flat plain, the shoreline will be broad and level. If it is located in the mountains, the shoreline is likely to be steep and rocky. Beaches that develop along rocky shorelines usually consist of pebbles and larger stones, as the waves carry away finer particles.

Movement of Earth's crust during an earthquake may have given the shoreline a folded appearance. Huge boulders and rocks or piles of gravel and sand may indicate that a glacier once moved across the region. The presence of a river may mean a large quantity of sedimentary deposits on the lake bed or the shoreline near the river mouth. Sandy shores are constantly changing, depending upon the movement of the wind, water, and sand. Some sandy shores may be steep while others have a gentle slope.

Landforms   Landforms include cliffs and rock formations; beaches and dunes; spits, bars, and shoals; deltas; and islands.

Cliffs and rock formations   Where highlands meet the edge of a lake, cliffs sometimes occur. Waves pounding the cliffs gradually eat into their base, creating a hollowed-out notch. The overhang of the notch may collapse and fall, creating a platform of rubble.

Many shorelines consist of both hard and soft rock. Wave action erodes the soft rock first, sometimes sculpting beautiful shapes, such as arches, along the shore. Caves may be gradually carved into the sides of cliffs, or headlands may be created. A headland is a large arm of land made of hard rock that juts out from shore after softer rock on either side has been eroded away.

Beaches and dunes   Beaches are nearly level stretches of land along the water's edge. They may be covered by sand or stones. Sand is small particles of rock, less than 0.08 inches (2 millimeters) in diameter. It may be white, golden, brown, or black, depending upon the color of the original rock. Yellow sand usually comes from quartz and black sand from volcanic rock. White sand may have been formed from limestone.

Putting a Nuisance to Work

A water hyacinth is an aquatic plant that lives in ponds, quiet streams, and ditches. Water hyacinths have large violet flowers that float on top of the water, and some species have long, trailing, feathery roots. In the wrong place, these roots can clog otherwise navigable watercourses and irrigation ditches, which makes the plant a nuisance. The water hyacinth may be able to make up for this annoying tendency with another important characteristic: it can absorb large amounts of pollutants. Researchers are experimenting with using the hyacinth in water-treatment systems and sewage-disposal plants.

Sand is carried by water and wind. When enough sand has been heaped up to create a ridge or hill, it is called a dune. Individual dunes often travel, as the wind changes their position and shape. They tend to shift less if grasses and other plants take root in them and help hold them in place.

Spits, bars, and shoals A spit is a long narrow point or strip of deposited sand, mud, or gravel that extends into the water. A bar is an underwater ridge of sand or gravel formed by currents that extends across a channel. Shoals are areas where enough sediments have accumulated that the water is very shallow. Shoals and bars make navigation dangerous for boats.

Deltas Where rivers meet a lake, huge amounts of silt (a type of very small soil particle) can be carried by river currents from far away and deposited in the lake along the shoreline. Large rivers can dump so much silt that islands of mud build up forming a fan-shaped area called a delta. The finer debris that does not settle as quickly may drift around the lake, making the water cloudy.

Large rivers deposit silt as the water slows down, entering a lake. This causes mud to build up, creating islands in a fan shape. IMAGE COPYRIGHT MIRCEA BEZERGHEANU, 2007. USED UNDER LICENSE FROM SHUTTERSTOCK.COM.

Islands   Islands are landmasses completely surrounded by water. In lake basins dug out by glaciers, islands may be formed from large piles of debris created when the glacier moved across the area. In crater lakes, islands may have formed from secondary cones that developed when the volcano became active again. Other islands may have been created when the water surrounded a high point of land, cutting it off from the shore on all sides.

Basins   Basins of ponds or lakes slope gently from the shore down to the center, where the water is deepest. Basins are usually covered with sediments. Some sediments are formed by waste products and dead tissues of plants and animals. Others consist of clay, stone, and other minerals.

Elevation   Lakes occur at all altitudes. For example, Lake Titicaca, which lies between Bolivia and Peru, is 12,500 feet (3,810 meters) above sea level, while the Sea of Galilee, located between Israel and Jordan, is 695 feet (212 meters) below sea level. (Sea level refers to the average height of the surface of the sea.)

Plant Life

Most lake and pond plants live in the waters around the shoreline. They include microscopic, one-celled organisms; plants commonly referred to as seaweed; and many other types of grasses and flowering plants.

The water offers support to plants. Even a small tree on land requires a tough, woody stem to hold it erect, but underwater plants do not require woody portions because the water helps to hold them upright. Their stems are soft and flexible, allowing them to move with the current without breaking.

Plants in and around a lake or pond may be classified as submergent, floating aquatic (water), or emergent, according to their relationship with the water.

A submergent plant grows beneath the water. Even its leaves are below the surface. Submergents include milfoil, pondweed, and bladderwort, an insect-eating plant.

Floating aquatics float on the water's surface. Some, such as the water hyacinth, water lettuce, and duckweed have no roots anchored in the bottom soil. Others, such as water lilies and pondweed, have leaves that float on the surface, stems that are underwater, and roots that anchor them to the bottom.

Water lilies and lily pads are common lake and water plants.
COPYRIGHT © 2006 KELLY A. QUIN.

An emergent plant grows partly in and partly out of the water. The roots are usually underwater, but the stems and leaves are at least partially exposed to air. They have narrow, broad leaves, and some even produce flowers. Emergents include reeds, rushes, grasses, cattails, and water plantain.

Plants and plantlike organisms that live in lakes and ponds can be divided into four main groups: algae (AL-jee), fungi (FUN-ji), lichens (LY-kens), and green plants.

Algae   Most lake and pond plants are algae. (It is generally recognized that algae do not fit neatly into the plant category.) Algae inhabit even the salt and soda lakes that are unfriendly to other life forms.

Some forms of algae are so tiny they cannot be seen without the help of a microscope. Other species, like many seaweeds, are larger and remain anchored to the bed of the lake or pond.

Growing season   Algae contain chlorophyll, a green substance used to turn energy from the sun into food. As long as light is available, algae can grow. Growth is often seasonal. In some areas, such as the northern hemisphere, the most growth occurs during the summer when the sun is more directly overhead. In temperate (moderate) zones, growth peaks in the spring but continues throughout the summer. In regions near the equator, no growth peaks occur. Instead, growth is steady throughout the year.

Food   Most algae have the ability to make their own food by means of photosynthesis (foh-toh-SIN-thih-sihs). Photosynthesis is the process by which plants use light energy to change water and carbon dioxide from the air into the sugars and starches they use for food. A by-product of photosynthesis is oxygen, which combines with water and enables aquatic animals, such as fish, to breathe. These types of algae grow in the euphotic, or sunlit zone where light is available for photosynthesis. Other types absorb nutrients from their surroundings.

Algae require other nutrients that must be found in the water, such as nitrogen, phosphorus, and silicon. In some lakes, when deeper waters rise during different seasons and mix with shallower waters, more of these nutrients are brought to the surface. Algal growth increases when nitrogen and phosphorus are added to a body of water by sewage or by runoffs from farmland.

Reproduction   Algae reproduce in one of three ways. Some split into two or more parts, each part becoming a new, separate plant. Others form spores (single cells that have the ability to grow into a new organism). A few reproduce sexually, during which cells from two different plants unite and a new plant is created.

Common lake and pond algae   Two types of algae are commonly found in lakes and ponds: phytoplankton and macrophytic algae.

Phytoplankton float on the surface of the water, always within the sunlit zone. Two forms of phytoplankton, diatoms and dinoflagellates (dee-noh-FLAJ-uh-lates), are the most common. Diatoms have simple, geometric shapes and hard, glasslike cell walls. They can live in colder regions and even within arctic ice. Dinoflagellates have two whiplike attachments that make a swirling motion. They often live in tropical regions (regions around the equator).

Macrophytic algae (*macro* means large) usually grow attached to the bottom of the lake or pond and are submergents.

Fungi   As with algae, fungi and lichens do not fit neatly into the plant category. Fungi are plantlike organisms that cannot make their own food by means of photosynthesis. Instead, they grow on decaying organic (derived from living organisms) matter or live as parasites on a host. A parasite is an organism that depends upon another organism for its food or other needs. Fungi grow best in a damp environment, which makes the edges of lakes and ponds a favorable home. Common fungi include mushrooms, rusts, and puffballs.

Lichens   Lichens are combinations of algae and fungi that tend to grow on rocks and other smooth surfaces. The algae produces food for both itself and the fungus by means of photosynthesis. In turn, it is believed the fungus protects the algae from dry conditions.

Green plants   Hydrophytes are green plants found growing in the water or in very wet places where the soil is saturated (soaked with water). They

Waterproof Roofing

Reeds have been used for thousands of years as a building material. During the Middle Ages (500–1500) in England, for example, people used reeds to make thatched roofs for their houses. The reeds are bundled tightly together and, when put in place, keep out both rain and cold. If made by a skilled worker with reeds of good quality, a thatched roof will remain waterproof for up to forty years.

Mexico's Floating Gardens

The Aztecs, the original settlers of Mexico, built rafts that they floated on Lake Xochimilco (soh-chee-MEEL-koh), which lies about 12 miles (19 kilometers) southeast of Mexico City. The rafts were used to grow flowers, vegetables, and fruits. But because the lake was shallow, the rafts soon became rooted in the water. Eventually a city formed, and today Xochimilco is famous for these permanent floating gardens.

are similar to plants that grow on dry land and are found along the shoreline of lakes and ponds. Sedges are an example. They have roots adapted to this environment. Large beds of these plants slow the movement of water and help prevent erosion of the shoreline. Some water animals use them for food and hiding places.

Mesophytes such as reeds, need moist, but not saturated soil. They occupy the emergent zone between the water-covered area and dry land. The plants that grow on the dry shore are called xerophytes. They are able to survive with little moisture and are more typical of dry, arid lands.

Most green plants need several basic things to grow: light, air, water, warmth, and nutrients. Light and water are in plentiful supply near a lake or pond. Nutrients, primarily nitrogen, phosphorus, and potassium, are usually obtained from the soil. Some soils are lower in these nutrients. They may also be low in oxygen so many lake and pond plants have special tissues with air pockets that help them to breathe.

Common lake and pond green plants Typical green plants found around lakes and ponds include water lilies, pondweed, and duckweed.

Water lilies are found in both temperate and tropical regions. They have large, nearly circular leaves that float on the water and beautiful white, yellow, pink, red, scarlet, blue, or purple flowers that float or are supported by stems above the water's surface. Ancient peoples considered the water lily a symbol of immortality because it arose from dried-up pond beds after the return of the rains.

Pondweeds usually have both floating and submerged leaves, although some species are completely submerged. The flowers appear above the water on spikes and bear a nutlike fruit. Pondweeds are important food plants for ducks.

Growing season Climate and precipitation affect the length of the growing season. Warmer temperatures and moisture usually signify the beginning of growth. In regions that are cold or receive little rainfall, the growing season is short. Growing conditions are also affected by the amount of moisture in the soil, which ranges from saturated to dry.

Reproduction   Green plants reproduce by several methods. One is pollination, in which the pollen from the male reproductive part of a plant, called the stamen in flowering plants, is carried by wind or insects to the female reproductive part of a plant, called a pistil in flowering plants. Pollen-gathering insects distribute pollen at different times of the day according to when a plant's flower is open. Water lilies, for example, which are closed in the morning and evening, open during midday when the weather is warmer and insects are more active. Instead of pollination, some shoreline plants grow rhizomes, which are stems that spread out under water or soil and form new plants. Reed mace and common reeds develop underground rhizomes.

Duckweeds are small water plants that may be the size of a grain of rice that float on the surface during spring and summer. During the growing seasons the plants produce an excess of starch that weighs them down, so that by autumn they sink to the bottom. The starch keeps them alive throughout the cold months, and in the spring they float to the surface again.

Snowy white egret looks for fish in Florida wetland marsh pond covered with duckweed. IMAGE COPYRIGHT FLORIDASTOCK, 2007. USED UNDER LICENSE FROM SHUTTERSTOCK.COM.

Endangered species   Changes in the habitat, such as succession, pollution, and new weather patterns, endanger lake and pond plants. Lakes and ponds are endangered by people who collect the pond plants, and who use insecticides (insect poisons) and herbicides (plant poisons) on farms and gardens. These poisons contaminate nearby rivers and streams, which then contaminate the lake or pond. Herbicides in particular are harmful to plant life. Other threats include fertilizer runoff from nearby farms. Saw grass, for example, struggles to survive in water polluted by fertilizer.

Animal Life

Freshwater lakes and ponds support many species of aquatic and land animals. Some aquatic animals swim freely in the water; others live along the bottom. Many land animals, such as opossums, raccoons, and deer, visit lakes and ponds for food and water.

Water Walkers

What allows some insects, such as the water strider, to skate along the surface of the water without sinking? A water strider is so small that its body is extremely light. Within a lake or pond, each molecule of water has similar molecules attracting it uniformly in every direction. At the water's surface, these molecules crowd together to produce what is called surface tension. If a creature or an object is small and light enough, these tightly packed molecules will support its weight. A steel sewing needle, for example, can float on water if it is gently lowered to rest on the surface.

Few species of animals live in the waters of salt and soda lakes. Those that do provide food for birds and other animals. Ducks and wading birds feed on the microscopic brine shrimp that flourish in these lakes. These types of lakes are usually found in desert areas and they may be the only source of food for many miles.

Microorganisms Microorganisms cannot be seen by the human eye without the use of a microscope. Those found in ponds and lakes include transparent daphnia, creatures that catch food with the hairs on their legs; rotifers that sweep algae into their mouths with swirling hair-like fibers; and bacteria. Bacteria make up a large portion of the microscopic organisms found in lakes and ponds. They aid in the decomposition (breaking down) of dead organisms. Bacteria tend to exist in great numbers near the shoreline where larger organisms are found.

Invertebrates Invertebrates are animals without a backbone. They range from simple worms to more complex animals such as insects and crabs.

Many species of insects live in lakes and ponds. Some, like the diving beetle, spend their entire lives in the water. Others, like mosquitoes, live in the water as larvae but are able to breathe air through small tubes attached at the end of their bodies. They leave the water when they become adults. Another type of insects, like backswimmers, have gills, just like fish, which enable them to obtain oxygen from the water.

Crustaceans, such as shrimp, and mollusks, such as mussels, are invertebrates with a hard outer shell that often inhabit lakes and ponds. One species, the brine shrimp, favors salt and soda lakes where it is able to filter out the harmful minerals.

Leeches feed on blood, and as they feed they produce a blood-thinning chemical called hirudin. Hirudin is used for medical purposes, and about 26,000 pounds (57,200 kilograms) of leeches are caught each year commercially. Their numbers have diminished as a result.

Common lake and pond invertebrates Invertebrates common to lakes and ponds include fisher spiders, leeches, and dragonflies.

Fisher spiders are predators that feed on insects and tadpoles. The fisher's body is covered with hairs that help distribute its weight and allow it to walk on water. If the spider is submerged, the hairs trap a coating of air so the animal can remain underwater for a long period of time and still breathe.

Leeches are slick, flat parasites that live on the blood of other animals. Some attack fish, and others attack snails, reptiles, and mammals, including humans. Turtles, for example, often have a ring of leeches around their eyes and necks. As it feeds, the body of the leech swells with blood. Many species feed only occasionally because they can store food in their digestive systems.

Water spiders can stay suspended over water, aided by the air bubbles under their feet.
IMAGE COPYRIGHT
BASEL101658, 2007. USED
UNDER LICENSE FROM
SHUTTERSTOCK.COM.

Dragonfly nymphs, the youthful form of the adult insect, are among a lake or pond's dominant carnivores (meat eaters). They will eat anything smaller than themselves, including fish, snails, and other insects. Some stalk their prey; others lie in wait and nab it as it passes. The nymph grasps the victim with its lower lip and pulls it into its mouth.

Food   Some insects feed underwater as well as on the surface. They may be either plant eaters, meat eaters, or scavengers that eat decaying matter. The giant water bug is a 3-inch-long (76-millimeter-long) predator that grasps fish, frogs, and other insects in its powerful legs, paralyzes them with injections of poison, and sucks out their body fluids.

The diets of other invertebrates also vary. Some snails are plant feeders that eat algae, whereas crabs are often omnivorous, eating both plants and animals.

Reproduction   Most invertebrates that are insects have a four-part life cycle. The first stage is spent as an egg. The second stage is the larva, which may actually be divided into several stages as the larva increases in size and sheds layers of the outer skin. The third stage is the pupal stage, during which the insect lives in yet another protective casing. The pupal stage is the final stage of development before emerging as an adult.

Amphibians   Amphibians, including frogs, toads, newts, and salamanders, are vertebrates, which means they have a backbone. Amphibians live at least part of their lives in water. They must usually remain close to a

Bullfrogs are the largest North American frog. IMAGE COPYRIGHT PAUL S. WOLF, 2007. USED UNDER LICENSE FROM SHUTTERSTOCK.COM.

water source because they breathe through their skin, and only moist skin can absorb oxygen. If they are dry for too long, they will die.

Amphibians are cold-blooded animals, which means their body temperatures are about the same temperature as their environment. They need the warmth they get from the sun in order to be active. As temperatures grow cooler, they slow down and seek shelter. In cold or temperate regions, some amphibians hibernate (remain inactive) and dig themselves into the mud. When the weather gets too hot, they go through a similar period of inactivity called estivation.

Common lake and pond amphibians The most common lake and pond amphibians are salamanders, newts, frogs, and toads. They can be found all over the world, primarily in fresh water.

The bullfrog is the largest North American frog. Its body measures 8 inches (20 centimeters) in length and its legs another 10 inches (25 centimeters). Long legs enable it to leap 15 feet (4.6 meters). A bullfrog spends most of its time floating in the water or diving to the bottom in search of food. It eats insects, crayfish, smaller frogs, small birds, and small mammals.

Food In their larval form, amphibians are usually herbivorous (plant eating). Adults are usually carnivorous, feeding on insects, slugs, and worms. Those that live part of their lives on land have long, sticky tongues with which they capture food.

Reproduction   Most amphibians lay jellylike eggs in the water. Depending on the species, frogs can lay as many as 50,000 eggs, which float beneath the water's surface. Frog eggs hatch into tadpoles (larvae) that can swim, and breathe through gills. Spotted newts hatch in the water and live there as larvae, developing gills. Later, they lose their gills and live for a time on dry land. Two or three years later, they return to the water where they live the remainder of their lives.

Reptiles   Reptiles are cold-blooded vertebrates that depend on the temperature of their environment for warmth. They are more active when the weather and water temperature become warmer. Many species of reptiles, including snakes, lizards, turtles, alligators, and crocodiles, live in temperate and tropical lakes.

Many reptiles go through a period of hibernation in cold weather because they are so sensitive to the temperature of the environment. Turtles, for example, bury themselves in the mud. They barely breathe and their energy comes from stored body fat. When the weather becomes very hot and dry, some reptiles go through estivation, another inactive period similar to hibernation.

Common lake and pond reptiles   Two well-known lake and pond reptiles are the grass snake and the painted turtle.

The grass snake is found in Europe, western Asia, and North Africa. Averaging a length of about 28-47 inches (70-120 centimeters), its diet consists primarily of frogs, newts, and fish. Although they do not spend all of their time in water, grass snakes are good swimmers and obtain most of their food from lakes and ponds.

The painted turtle usually lives in shallow, muddy, freshwaters of North America. It grows to between 3.5 and 10 inches (9 and 25 centimeters) and can be identified by the red and yellow markings on their dark colored shell.

Food   All snakes are carnivores. Many that live around lakes and ponds, including the banded water snake, eat frogs, small fish, and crayfish. The pond turtle is an omnivore. When it is young it feeds on insects, crustaceans, mollusks, and tadpoles. Adult turtles eat primarily wetland plants.

Reproduction   The shells of lizard, alligator, and turtle eggs are either hard or rubbery, depending upon the species, and do not dry out easily. Most species bury their eggs in warm ground, which helps them hatch. The eggs of a few lizards and snakes remain inside the female's body until they hatch, and the females give birth to live young.

The Monster of Loch Ness

Since the Middle Ages (500–1500), people have reported seeing a huge, serpentlike monster swimming in Loch Ness, a deep lake (754 feet; 230 meters) in northern Scotland, and the largest lake in Great Britain. The first recorded sighting was made in 565 when Saint Columba came upon the burial of a man said to have been bitten to death by the monster. According to one account, Saint Columba himself saw the creature.

In 1933, the monster, referred to as Nessie, attracted the attention of the news media when a man and woman driving by the lake noticed a great commotion of water in the middle, and for several minutes they watched "an enormous animal rolling and plunging." The incident was widely reported.

Since then, many people have tried to find evidence of the creature using such equipment as sonar (sound vibration technology) and even a submarine. In 1972 and 1975, an American expedition sponsored by the Academy of Applied Science obtained startling underwater time-lapse pictures that some researchers believe show a large animal swimming submerged in Loch Ness. A 1987 British expedition failed to confirm the presence of such a creature, and most scientists do not believe Nessie exists.

Crocodiles and alligators keep their eggs warm by laying them in nests, which can be simple holes in the ground or constructions above the ground made from leaves and branches.

Fish Fish are cold-blooded vertebrates. Lakes and ponds support two basic types. The first grouping are parasites and include lampreys, which attach themselves with suckers to other animals and suck their blood for food. The second grouping, the bony fishes, are the most numerous. They use fins for swimming and gills for breathing. Some fish can survive in polluted waters. Others, such as some African and Asian species of catfish, can breathe air and may live for a short time on land.

Many species of fish, such as sticklebacks, swim in schools. A school is a group of fish that swim together in a coordinated manner. The purpose of a school is to discourage predators. Also, the more eyes watching, the more likely it is that a predator will be seen before it can strike.

In the winter, most lakes freeze only on the surface. This surface ice helps insulate the deeper water by trapping the warmth, and allowing fish down below to survive. The ice is comparatively thin, so light can still reach submerged plants and they can photosynthesize, which helps put oxygen into the water, enabling the fish to breathe.

Common lake and pond fish   Typical lake and pond fish include the pike and the carp.

Pike are long, narrow, streamlined fish that live in shallow lakes and large ponds. They may grow as large as 80 pounds (36 kilograms), depending on the size of their habitat. The largest type of pike is the muskellunge, which is found in North American waters. Other species live in Europe and Russia.

Pike are predators and hide among water plants where their striped and spotted bodies blend in with the surroundings. They move swiftly and catch prey in their sharp teeth. When young, they feed on insects and worms, but large adults will attack water birds and small mammals.

Carp prefer warm, weedy water and are found across Europe, North America, South America, and parts of Asia, Africa, Australia, and New Zealand. The world's largest known carp, the giant Siamese carp, weighed 265 pounds (120 kilograms) and was caught in the lakes of Thailand.

Carp are resistant to pollution and can survive where other fish die. When young, they eat insect larvae and crustaceans. As adults, they eat invertebrates and aquatic plants.

Food   Some fish eat plants while others depend upon insects, worms, and crustaceans (shellfish). Larger fish often eat smaller fish, and a few species feed on carrion (dead bodies). Most fish specialize in what they eat. Bluegills, for example, feed on insect larvae, and pumpkin-seed fish eat snails. Some fish feed on the surface, and others seek food in deep water.

Reproduction   Most fish lay eggs. Many species abandon the eggs once they are laid. Others build nests and care for the new offspring. Still others carry the eggs with them until they hatch, usually in a special body cavity, or in their mouths.

Birds   Many different species of birds live on or near lakes and ponds. These include many varieties of wading birds, waterfowl, and shore birds. Most visit lakes or ponds in search of food, and certain species use them as nesting places.

Common lake and pond birds   Birds found around lakes and ponds can be grouped as wading birds, shore birds, waterfowl, or birds of prey.

Wading birds, such as herons, have long legs for wading through shallow water. They have wide feet, long necks, and long bills that are used for nabbing fish, snakes, and other food. Herons are common in

Swans are well adapted for swimming because their feet are close to the rear of their bodies, and their feet are webbed.
IMAGE COPYRIGHT OTMAR SMIT, 2007. USED UNDER LICENSE FROM SHUTTERSTOCK.COM.

North America, Europe, and Australia, and there are about 100 species worldwide.

Shore birds, such as the plover or the sandpiper, feed and nest along the banks of lakes and ponds. They prefer shallow water. The ruddy turnstone, a stocky shore bird with orange legs, is named for its habit of overturning pebbles and shells in search of food.

Waterfowl are birds that spends most of their time on water, especially swimming birds such as ducks, geese, or swans. Their feet are close to the rear of their bodies, and the skin between their toes is webbed. This is good for swimming but awkward for walking and causes them to waddle. Their bills are designed for grabbing the vegetation, such as sedges and grasses, on which they feed. There are approximately 150 species of these types of birds.

The fish eagles, which are birds of prey, hunt exclusively in freshwater. The bald eagle, the national bird of the United States, occasionally hunts for fish, as do kingfishers and osprey.

Food   Nearly all birds must visit a source of fresh water each day to drink, and many feed on aquatic vegetation or the animals that live in lakes and ponds. An adaptation of birds to aquatic life is their beaks, or bills. Some beaks, like that of the kingfisher, are shaped like daggers for stabbing prey such as frogs and fish. Others have a slender bill, like the least sandpiper, designed to probe through the mud in search of food, such as insect larvae and small mollusks.

Reproduction   All birds reproduce by laying eggs. Male birds are typically brightly colored to attract the attention of females. After mating, female birds lay their eggs in a variety of places and in nests made out of many different materials. Different species of birds lay varying numbers of eggs. The mute swan lays five to seven eggs, which hatch in about thirty-six days. One parent usually sits on the nest to keep the eggs warm.

Moose feed on twigs, bark, and saplings, as well as aquatic plants. IMAGE COPYRIGHT CALEB FOSTER, 2007. USED UNDER LICENSE FROM SHUTTERSTOCK.COM.

Mammals   Mammals are warm-blooded vertebrates covered with at least some hair, they bear live young, and nurse with the mother's milk. Aquatic mammals, such as muskrats, have waterproof fur and partially webbed toes for better swimming. Other mammals, such as raccoons, often visit lakes and ponds for food or water but spend little time in the water.

Common lake and pond mammals   Although many mammals visit lakes and ponds for food and water, some species spend most of their time in the water. These include muskrats, otters, water shrews, water voles, beavers, and, in summer, even moose.

The water vole is a mouselike rodent that is not a good swimmer but an excellent diver. It lives in burrows dug in the bank of a lake or pond. It eats mainly reeds, sedges, and other shoreline plants, as well as acorns and beechnuts. Nuts are usually stored for use in winter because the vole does not hibernate. The water vole is found in Europe and parts of Russia, Siberia, Asia Minor, and Iran.

At about 4 feet (1.3 meters) long, the beaver can weigh as much as 65 pounds (30 kilograms). The long, reddish-brown fur of the beaver is warm and waterproof, allowing it to swim in icy water. The toes on the hind feet are webbed for swimming, and its tail is shaped like a flat paddle, which helps it maneuver through the water. The front feet resemble small hands, enabling the beaver to carry things. Its large front teeth allow it to cut down trees and other vegetation. Beavers feed mainly on bark from aspen, willow, poplar, birch, and maple trees.

Beavers are social animals who live and work in groups. They live in lodges, a network of tunnels and burrows deep within log structures

Beavers are very industrious animals, and can be seen working on their dams that they use for shelter. IMAGE COPYRIGHT JOSEPH DIGRAZIA, 2007. USED UNDER LICENSE FROM SHUTTERSTOCK.COM.

called beaver dams. Beavers build these dams by cutting down small trees with their sharp teeth. They then pile logs and sticks across small rivers and streams, damming them up and creating ponds. The lodge usually has an underwater entrance and air holes for ventilation.

The North American beaver once ranged over the continent from Mexico to Arctic regions. It was widely hunted for its fur and for a liquid called castorium, produced in the beaver's musk glands and used in perfume. As a result, beaver numbers are now greatly reduced, and it is confined largely to northern wooded regions. Beavers were once common throughout northern Europe but are virtually extinct there, except in some parts of Scandinavia, Germany, and Siberia.

The moose is the largest member of the deer family. The largest on record weighed 1,166 pounds (528 kilograms). Moose antlers may span more than 6 feet (2 meters). They feed on twigs, bark, and saplings in winter, and they spend many hours in lakes and ponds in warm weather where they eat aquatic vegetation. They will even submerge completely to get at the roots of water plants.

Moose are found in wooded regions of North America. In other countries they are called elk (a close relative) and may be found in Norway, Sweden, Russia, and northern China.

Food Certain aquatic mammals, like the otter, are carnivores, eating rabbits, birds, and fish. Muskrats are omnivores, and they eat both animals, such as mussels, and plants, such as cattails. Beavers are herbivorous and eat trees, weeds, and other plants.

Reproduction　Mammals give birth to live young that have developed inside the mother's body. Some mammals are helpless at birth, while others are able to walk and even run almost immediately. Many are born with fur and with their eyes and ears open. Others, like the muskrat, are born hairless and blind.

Endangered species　As lakes and ponds are filled in or polluted, many species of migratory birds are threatened because they can no longer use them for finding food or as nesting areas. Their numbers decline as a result.

Human Life

Since prehistoric times, lakes and ponds have played an important role in the lives of the people who lived in the surrounding region.

Impact of lakes and ponds on human life　People use lakes and ponds for water and food, for recreation and building sites, for transportation, and for industrial purposes.

Water　Many lakes, such as Lake Michigan, are sources of drinking water, as well as water for such things as bathing, laundry, and power generation for nearby communities. The world's use of water has tripled since 1950 and in the United States alone, each person uses an average of 130 gallons of fresh water each day. In developed countries, about half the supply is used by industry. In less wealthy countries, 90 percent is used for crops and irrigation.

Food　The fish and plants in lakes and ponds are often a source of food for humans. Most fish used for food come from the ocean, but commercially important freshwater fish include catfish, lake trout, bass, perch, and whitefish. Some aquatic plant species, such as water chestnuts and watercress, are also popular foods. Farmers often use aquatic plants, such as marsh grass and sedges, for feeding livestock.

Recreation and building sites　Lakes and ponds are popular sites for sport fishing, swimming, boating, and nature appreciation. During

Prehistoric Lake Cabins

During several prehistoric periods, early humans often built homes and villages near the waters of a lake or a marsh. Probably in anticipation of flooding, the homes were built on platforms or artificial mounds. The most famous platform lake dwellings are those of the late Neolithic and early Bronze Ages (approximately 6,000 years ago) found in Switzerland, France, and northern Italy. One village contained about ninety circular huts constructed of close-set vertical timbers. Individual homes called crannogs were built on artificial mounds or islands in certain parts of Ireland and Scotland.

Hook, Line, and Sinker

Ever since the first humans tried to snatch minnows (tiny fish) from a pond, fishing has not only been a source of food, but also a popular sport. Ancient Egyptian drawings show fishing scenes, and the sport is mentioned in writings from China, Greece, Rome, and the Middle East.

In the United States alone, about 36.5 million freshwater fishing licenses are issued each year. The largest freshwater sport fish is the white sturgeon. Other popular freshwater fish include black bass, various types of trout, sunfish, crappies, salmon, perch, pike, muskies, sturgeon, and shad.

warm months, many lakes are crowded with vacationers, and in urban areas their shorelines have become popular sites on which to build homes. Artificial lakes or ponds may be created to add beauty and wildlife to a residential area.

Transportation   Large lakes, such as the Great Lakes in North America, provide water transportation. The Great Lakes were a way for settlers to travel to the interior of the North American continent. They are still important to commercial vessels and even ocean-going ships, because they are linked to the ocean by the St. Lawrence Seaway. Smaller lakes may be important for local boat traffic.

Other resources   Lake and pond plant materials can be used for building. Reeds are used for huts in Egypt and in stilt houses in Indonesia. Sediments, such as clay and mud, may be used to manufacture bricks.

Lake and pond fish provide more than just food, yielding such products as fish oils, fish meal, fertilizers, and glue.

Shoreline properties are popular places for people to live and relax. IMAGE COPYRIGHT THOMAS BARRAT, 2007. USED UNDER LICENSE FROM SHUTTERSTOCK.COM.

The water in lakes and ponds may be used to cool power stations and for other industrial purposes. Some industries have used lakes and ponds as dumpsites for industrial chemicals causing serious environmental damage.

Impact of human life on lakes and ponds   The important role of lakes and ponds for plant and animal communities, for human life, and for the environment has not always been appreciated. Many ponds have been drained to eliminate mosquitoes and prevent diseases, to increase land for farming, and to make room for development. The water in others has been carelessly used for irrigation or industrial purposes.

Water supply   Although all water on Earth is in constant circulation through evaporation and precipitation, some regions may have a limited supply. As populations grow, the supply diminishes even more. In some regions, forests are cut down and replaced by farms. Since trees help conserve underground water, much of this water may be lost. Ponds gradually disappear because they are often supplied by groundwater, springs, and lakes.

Irrigation practices can cause damage, especially in desert regions, because there is not enough precipitation to replace the water used to water crops. The Aral Sea, a large lake in Central Asia has shrunk to half its original surface area because so much water has been removed to irrigate farms in the area.

Use of plants and animals   Many freshwater lakes used for recreation have been overfished, and attempts to stock the lakes with commercially raised fish are not always successful. Overfishing is often more serious in lakes where the local population depends upon fish for their daily food. The fish are rarely restocked and their numbers never recover.

Overdevelopment   People may wish to live near a lake or pond for its beauty and a sense of being close to nature, but overdevelopment can destroy the very things they are seeking. Overdevelopment results in erosion of the shoreline and loss of the scenic value. Instead of blue water and trees, residents look out on cars and concrete.

Quality of the environment   Pollution by communities, industries, shipping, and poor farming practices has led to poisoning of water and changes in its temperatures. Several environmental problems, including accelerated eutrophication (YOU-troh-fih-kay-shun; becoming over enriched with nutrients) and acid rain, are issues that affect the quality of water in lakes and ponds.

Lakes of the Poets

The Lake District in northwestern England is a region of mountains, lakes, and waterfalls famous not only for its beauty but also for the many great poets it inspired. William Wordsworth (1770–1850), Samuel Taylor Coleridge (1772–1834), and Robert Southey (1774–1843) all lived in the district.

The Lake District National Park, established in 1951, covers an area of 866 square miles (2,243 square kilometers) and takes in England's largest lakes and highest mountains, including Scafell Pike. Lakes include Lake Windermere, Ullswater, Bassenthwaite Lake, Derwent Water, and Coniston Water.

Eutrophication occurs when fertilizers, especially nitrogen and phosphorus, used in farming get into lakes and ponds, spurring a greatly increased growth of algae. These algae form a thick mat on the water's surface and block the sunlight, causing submerged plants to die. As the dead plants decay, oxygen in the water decreases. Water with little oxygen cannot support most plants and animals. Some of the world's major lakes, such as Lake Geneva in Switzerland, suffer from eutrophication. Eutrophication results in the lake becoming a wetland (an area where the soil is saturated with water for most of the year).

Acid rain is a type of air pollution dangerous to lakes and ponds. It forms when industrial pollutants, such as sulfur or nitrogen, combine with moisture in the atmosphere to form sulfuric or nitric acids. These acids can be carried long distances by the wind before they fall, either as dry deposits or as rain or snow. Acid rain can damage both plant and animal life. It is especially devastating to amphibians such as salamanders because their skin is so thin and permeable (able to pass through). The pollutants from acid rain pass right through the skin and enter into their bodies.

The Ramsar Convention   The U.S. Department of the Interior, the U.S. Fish and Wildlife Service, and many environmental groups are working to preserve lakes and ponds and to create new ones. The Convention on Internationally Important Wetlands was signed by representatives of many nations in Ramsar, Iran, in 1971. Commonly known as the Ramsar Convention, it is an intergovernmental treaty that provides the framework for national action and international cooperation for the conservation and wise use of wetlands and their resources.

Native peoples   Humans have always lived near lakes and ponds because they provide a source of food and fresh water. Two groups that continue to depend on lakes for their survival are the Turkana in Kenya and native peoples of the Andes Mountains in South America.

The Turkana   The Turkana people live along the shores of Lake Rudolf in Kenya, renamed Lake Turkana. It covers an area of 2,433 square miles (6,405 square kilometers) and supports many fish species.

The Turkana did not always depend upon the lake for their livelihood. Until the 1960s they were nomads and raised herds of camels, goats, and cattle that they used for milk and blood, their primary foods.

In the 1960s a severe drought (extremely dry period) struck the region and many animals died, causing famine among the Turkana. The people turned to fishing and discovered Lake Turkana to be a rich source of large Nile perch and a smaller fish called tilapia. This discovery changed the Turkana's lifestyle. They began to catch fish for food, some of which they sold in nearby towns and cities.

The Turkana are having to change their lifestyle again, because Lake Turkana is shrinking, becoming more saline (salty), and has been overfished. The region gets less rainfall than before, and at least one of the rivers that feed the lake has dried up. Many Turkana have gone back to raising animals, although this cannot be a permanent solution because the drier climate has reduced grazing land.

Island People of Lake Titicaca   Lake Titicaca (tee-tee-KAH-kah) is the highest navigable lake on Earth at 12,580 feet (3,834 meters). It is located in the Andes Mountains, on the border between Bolivia and Peru in South America. The Native American people who live there are descendants of the ancient and powerful Inca Empire, and the lake has

In Lake Titicaca in Peru, reeds grow together, forming floating islands that people live on.
IMAGE COPYRIGHT JOEL BLIT, 2007. USED UNDER LICENSE FROM SHUTTERSTOCK.COM.

been a center of their lives for 2,000 years. The Incas believed the lake was the origin of human life. The lake contains forty-one islands, the largest is Isla del Sol (Sun Island).

A species of reed called totora grows in the lake and forms floating islands on which the people live. They use the reeds to build boats and huts and to make baskets, which they sell. They use submerged plants, called yacco, to feed their cattle that graze along the shore. Around 1930, non-native fish, trout and mackerel, were introduced to the lake, reducing the number of native species of karachi and boga.

The Food Web

The transfer of energy from organism to organism forms a series called a food chain. All the possible feeding relationships that exist in a biome make up its food web. In lakes and ponds, as elsewhere, the food web consists of producers, consumers, and decomposers. These three types of organisms transfer energy within the biome.

Algae and plants are the primary producers in lakes and ponds. They produce organic materials from inorganic chemicals and outside sources of energy, primarily the sun.

Animals are consumers. Those that eat only plants, such as snails, are primary consumers in the lake or pond food web. Secondary consumers, such as certain fish, eat the plant-eaters. Tertiary consumers are the predators, like turtles and snakes. Humans fall into the predator category. Humans are omnivores, which means they eat both plants and animals.

Decomposers feed on dead, organic matter. These organisms convert dead organisms to simpler substances. Decomposers include insect larvae and bacteria.

Harmful to the lake or pond food web is the concentration of pollutants and dangerous organisms. They become trapped in sediments where certain life forms feed on them. These life forms are fed upon by other life forms, and at each step in the chain the pollutant becomes more concentrated. Finally, when humans and other large mammals eat fish, ducks, or other animals contaminated with pollutants, they are in danger of serious illness. The same is true of diseases such as cholera, hepatitis, and typhoid, which can survive and accumulate in certain aquatic animals and then be passed on to people who eat these animals.

Spotlight on Lakes and Ponds

Lake Baikal   Lake Baikal is the largest lake on Earth by volume, 5,521 cubic miles (23,015 cubic kilometers). It contains 20 percent of the world's fresh water and 80 percent of the fresh water in Russia and Siberia. It is also the world's deepest lake, reaching 5,712 feet (1,740 meters), with a layer of accumulated sediment on its floor adding another 5 miles (8 kilometers) of depth. Although Baikal has the largest surface area of any lake in Europe or Asia, it is small in comparison to the area covered by the Great Lakes in North America. More than 300 rivers flow into Lake Baikal, but only one, the Angara, flows out.

Lake Baikal formed during a shifting of the Earth's crust 25 million years ago, making it the world's oldest freshwater lake. Earthquake activity continues in the region, and some scientists think that the Asian continent is splitting apart at this point. If so, Baikal may actually be a "baby" ocean. This theory is supported by the 1990 discovery of hot water springs bubbling up through the floor, a feature usually found in mid-ocean.

Many storms develop over the surface of Baikal because it is so large; often whipping up waves higher than 15 feet (4.6 meters). The lake has a moderating effect on local weather. Its water temperature is cold, never climbing to more than 63°F (17°C) in summer. From January to May the lake is frozen to a depth of 3 feet (1 meters).

Baikal's water has few minerals. The presence of unique miniature crustaceans (shrimp and crab) that consume the millions of phytoplankton help to keep the lake very clear and blue. It is often called the blue pearl of Siberia.

Of the 1,500 known animal species and 700 plant species found in Baikal, two-thirds are found nowhere else on Earth. Most of these unique species are found in its deepest waters where light never penetrates and water temperature remains at about 32°F (0°C) for the year. Most of the creatures have little eyesight or are blind. One species of fish, the golomyanka, has no scales and contains so much oil in its body that it is translucent. The nerpa seal, one of only two species of freshwater seals, lives in Baikal. These seals feed primarily on fish, which feed on freshwater shrimp, which are being harmed by pollution. Thus, the seals' food supply is threatened.

At one time, Baikal was one of the cleanest lakes in the world and a popular summer resort. In the 1960s industrial waste began to threaten its

Lake Baikal
Location: Siberia
Area: 12,160 square miles
(31,494 square kilometers)
Classification: Fresh

purity. By the 1980s, the government had placed severe restrictions on activities that were polluting Baikal. In 1996, the area was deemed a UNESCO World Heritage site, meeting all four criteria: geological significance, biological-evolutionary importance, natural beauty, and outstanding importance for conservation.

The Great Lakes   The Great Lakes are a group of five freshwater lakes—Superior, Michigan, Huron, Erie, and Ontario—that lie along the border between the United States and Canada. Of the five lakes, Lake Superior is the largest and deepest. It covers a surface area of 31,800 square miles (82,362 square kilometers) and is, at its lowest point, 1,333 feet (406 meters) deep. The lakes are linked together by rivers and channels to create a waterway 2,342 miles (3,747 kilometers) long. Taken together, they comprise the largest surface area of fresh water in the world and the second largest in volume, 5,439 cubic miles (22,684 cubic kilometers).

The Great Lakes began to form about 60,000 years ago when glaciers moved across North America covering the region with ice. The weight of this ice depressed Earth's crust by as much as 1,500 feet (456 meters), and the movement of the ice carved out large basins in the rock. Glacial meltwater, combined with precipitation, gradually filled the basins, forming the lakes.

The lakes help moderate temperatures in the region, producing milder winters and cooler summers. In winter, the presence of the lakes contributes moisture to the air that helps produce lake-effect blizzards. Buffalo, New York, for example, receives an average 93 inches (236 centimeters) of snow a year. Sudden and severe storms are common. The Great Lakes Storm of 1913, often called the white hurricane, claimed eight large freighters and 235 lives.

In 1700, the Great Lakes supported many species of fish such as lake trout, sturgeon, and whitefish. By the end of the nineteenth century populations were greatly reduced, not only because of heavy fishing, but because of human tampering with the lakes' many tributaries (rivers and streams that feed lakes). Dams and destruction of surrounding forests destroyed spawning habitat. Ocean fish such as the alewife and the sea lamprey, which normally only lay their eggs in fresh water, have become permanent residents, competing for food with native species. In the 1960s, scientists introduced trout, coho salmon, and other predators that feed on lampreys, to help reduce the undesirable populations.

The Great Lakes
Location: North America
Area: 94,950 square miles (245,919 square kilometers)
Classification: Fresh

The sturgeon, which may be 6 feet (1.8 meters) in length, is prized for its meat and caviar. They have been overfished and are rarely found.

Connected by channels or canals, the lakes are navigable from Duluth, Minnesota, in the west to the eastern shore of Lake Ontario. Ocean-going vessels from the Atlantic Ocean can use the lakes by entering through the St. Lawrence Seaway. In 2002, 162 million tons (147 million metric tons) of dry bulk, such as iron ore, stone, coal, and salt, was moved on the lakes.

Fertilizer and farm runoff that has entered the lakes has increased the level of nutrients. This encourages rapid growth of phytoplankton and other plants, which make the lakes age faster. Industrial chemicals, such as dioxin and mercury, and other pollutants have destroyed the purity of the water, leading to destruction of plant and animal life.

Great Salt Lake　The Great Salt Lake lies between the Great Salt Lake Desert and the Wasatch Mountains in northwestern Utah. Most of the region is arid (dry). Approximately 4,200 feet (1,280 meters) above sea level, the lake was formed by glacier action. The Great Salt Lake is a remnant of a prehistoric freshwater lake ten times larger than it is now. Its size fluctuates, as does its depth, depending upon rates of evaporation.

The lake is fish free but some species of algae and a few invertebrates, such as brine shrimp and brine (alkali) flies, live in the lake. Many birds, including gulls, pelicans, blue herons, cormorants, and terns, nest on the lake's islets (tiny islands).

The Great Salt Lake is three to five times more saline (salty) than the ocean. Salts found in the lake include sodium chloride (table salt), and thousands of tons of this salt have been removed over the years for commercial purposes. The lake is used for recreation and has several beaches as well as a yacht harbor. Salt Lake City is built to the southeast and named for the lake.

Great Salt Lake
Location: Salt Lake City, Utah
Area: 2,450 square miles (6,345 square kilometers)
Classification: Salt

Mono Lake　Mono Lake lies in the shadow of the Sierra Nevada Mountains and was described by American writer Mark Twain (1835–1910) as "solemn, silent, sailless ... the lovely tenant of the loneliest spot on Earth." At least 700,000 years old, Mono was created by glaciers. All the water that enters it eventually evaporates, and for centuries it remained about the same size. Since 1941, many of the streams that normally feed it have been diverted to provide water for the city of Los Angeles. As a result, Mono Lake is shrinking, and the concentration of its minerals is

Mono Lake
Location: Eastern California
Area: 60 square miles (155 square kilometers)
Classification: Soda/salt

increasing. More than 25 square miles (65 square kilometers) of mineral-encrusted lake bottom have been exposed. Winds, picking up alkaline dust, create thick clouds thousands of feet (meters) in the air.

Mono's salinity (saltiness) is only three times greater than that of the oceans; however, its alkalinity is 80 times greater. Dissolved carbonates give it a bitter taste and create weird towers of mineral deposits, called tufa towers, that are exposed along the shoreline as the lake recedes.

Blue-green algae, brine shrimp, and brine (alkali) flies are among the few life forms Mono Lake supports. There are no other animals to eat the flies and shrimp, so they multiply unchecked. As many as 4,000 flies have been counted in a single square foot (.09 meter) of shoreline, and 50,000 brine shrimp have been found in a cubic yard (cubic meter) of lake water. The flies and shrimp attract many species of birds, including gulls, grebes, phalaropes, and ducks.

Lake Victoria   Lake Victoria, also called Victoria Nyanza, is the largest lake in Africa and the second largest freshwater lake in the world. Its coastline is more than 2,000 miles (3,220 kilometers) long, and it lies at an altitude of 3,720 feet (1,134 meters) above sea level. Its greatest depth is 270 feet (82 meters).

Victoria's basin was created by tectonic (earthquake) activity, and the shoreline is marked by steep cliffs, rocky headlands, swamps, and a river delta. Ukerewe and the Ssese Archipelago are its most important islands, all of which are densely wooded. The islands have become a popular destination for tourists.

The surrounding area has a large population, most of which are Bantu-speaking tribes. Boat building and fishing are important occupations. The introduction of Nile perch and Nile tilapia in the 1930s has caused at least 200 other species of fish to become extinct or near extinction. Lake Victoria yields about 550,000 tons (500,000 metric tons) of fish a year. This predominately includes Nile perch, Nile tilapia, and sardinelike dagaa.

A primary source of the Nile River, Lake Victoria was named for British monarch Queen Victoria (1837–1901) by British explorer John Hanning Speke (1827–1864), who first reached the lake in 1858. Alan Moorhead's books, *The White Nile* and *The Blue Nile*, and a British film titled *Mountains of the Moon*, tell of Speke's explorations.

Lake Victoria
Location: Africa, primarily within Uganda and Tanzania and bordering on Kenya
Area: 26,828 square miles (69,484 square kilometers)
Classification: Fresh

Caspian Sea The Caspian Sea is the world's largest lake in terms of surface area. It is bordered by Russia, Azerbaijan, Kazakhstan, Turkmenistan, and Iran. Lying 94 feet (28 meters) below sea level, its depth reaches 3,360 feet (1,024 meters) in the south. The depth in the northern portion of the sea is only 16 feet (5 meters). Its size fluctuates over time and, until 1977, the lake was shrinking.

The Caspian's basin was formed by both earthquake and glacier activity. Once part of an ocean, much of the surrounding region is covered by greenish clay ocean deposits. Its water is three times less salty than the ocean. The Ural and Volga rivers supply freshwater to the sea, diluting the salt content.

Average winter temperatures in the region are 14°F (-10°C) in the north and 50°F (10°C) in the south. In summer, the average is 79°F (26°C). Precipitation is light with only about 8 inches (20 centimeters) annually. In the shallower northern portion, ice forms during the winter months.

Over 400 species of fish live in the Caspian, including sturgeon, herring, pike, catfish, and carp. The sturgeon, comprising seven species in the sea, are fished for caviar. The Caspian seal, a freshwater seal, also makes its home here.

The lake is important to the economies of the surrounding regions, especially that of Russia and Iran. Many fish are taken from its waters and it is used for transportation. Fluctuations in its size and depth have reduced both the numbers of fish caught and the usefulness of its ports. Large oil fields extend beneath the lake, and there is considerable offshore production.

Lake Titicaca Located in the Andes Mountains at an altitude of 12,500 feet (3,810 meters), Lake Titicaca is the world's highest lake that can be used by large ships.

The presence of the lake moderates the local climate so that crops, such as corn, barley, quinoa (KEEN-wah), and potatoes, which are not usually grown at such high altitudes, can be raised there.

Only two species of fish, killifish, and catfish, inhabit the lake naturally. Trout were introduced in 1939.

The shores of the lake are densely populated by descendents of the Inca Indians. Modern steamboats and traditional Inca reed boats connect villages along the shoreline.

Caspian Sea
Location: On the border between Europe and Asia
Area: 144,000 square miles (373,000 square kilometers)
Classification: Salt

Lake Titicaca
Location: South America, between Bolivia and Peru
Area: 3,200 square miles (8,300 square kilometers)
Classification: Fresh

Dead Sea   The Dead Sea, a salt lake in the Middle East's Jordan Trench, has a shoreline located on the lowest point on Earth's surface—about 1,300 feet (396 meters) below sea level. At its deepest point, the lake descends to 2,300 feet (701 meters). Its basin was created by earthquake action, and at one time it was part of the Mediterranean Sea. Highlands border it to the east and west.

The region has a desert climate with warm, dry winters and hot, dry summers. Only about 4 inches (10 centimeters) of rain fall each year. The Dead Sea is fed by rivers but evaporates at the rate of about 55 inches (140 centimeters) each year. About nine times as salty as the ocean, its water is so dense that swimmers cannot sink and nothing grows there.

Minerals, such as potash (used in fertilizer), are mined from the lake, as are its salts. Muds from the shores of the sea are famous for their healing and rejuvenating effects.

The lake is a part of biblical history and figures in the stories of Abraham, Lot, David, Solomon, and the defenders of Masada. The first Dead Sea Scrolls (ancient Jewish manuscripts dating as far back as 350 BC) were found at Qumran on the northeastern shore.

Lake Clark   Lake Clark in southern Alaska is located about 125 miles (200 kilometers) southwest of Anchorage. It is part of Lake Clark National Park and Preserve, which is a mountainous region with two active volcanoes—Iliamna (10,016 feet/3,053 meters) and Redoubt (10,197 feet/3,108 meters). Gases are frequently seen venting out of Mount Iliamna, but there have been no recent eruptions. Mount Redoubt last erupted in 1966 and 1989.

Lake Clark is of glacial origin, and dozens of long valley glaciers, hundreds of waterfalls, and other glacial lakes are present in the park.

Lake Clark is the spawning ground for red salmon, and bald eagles and peregrine falcons live in the park year-round. The lake contains five different mammals: harbor seals, beluga whales, Stellar sea lions, and harbor porpoises. Large animals native to the park include grizzly bears, brown bears, wolves, lynxes, Dall sheep, moose, and caribou. The Dall sheep are the only wild sheep in the world with a white coat.

Crater Lake   Crater Lake, the deepest lake in the United States, is located in Crater Lake National Park in the Cascade Mountains of Oregon. The lake occupies the crater of an extinct volcano called Mount Mazama and lies at an altitude of 6,176 feet (1,882 meters). It is surrounded by lava

Dead Sea
Location: Between Israel and Jordan
Area: 405 square miles (1,049 square kilometers)
Classification: Salt

Lake Clark
Location: Alaska
Area: 110 square miles (286 square kilometers)
Classification: Fresh

Crater Lake
Location: Southern Oregon
Area: 20 square miles (52 square kilometers)
Classification: Fresh

walls as high as 2,000 feet (610 meters), and its depth is 1,943 feet (592 meters). Along the lake's western shore is Wizard Island, a small volcanic cone. Some geologists believe the volcano is not really extinct but only temporarily inactive.

Originally named Deep Blue Lake, Crater Lake glows intensely blue when the sun shines on it. Precipitation filled the lake and continues to maintain it. An average 44 feet (13.5 meters) of winter snow maintains the lake's water supply.

The surrounding area is heavily forested with lodgepole, ponderosa, and sugar pines, and white and Douglas firs. At higher elevations, stands of pine and fir are broken by meadows of wildflowers.

Eagles, hawks, owls, beavers, bears, mountain lions, and deer are common animals in the area.

For More Information

BOOKS

Day, Trevor. *Biomes of the Earth: Lakes and Rivers.* New York: Chelsea House, 2006.

Gleick, Peter H. et al. *The World's Water 2004-2005: The Biennial Report on Freshwater Sources.* Washington DC: Island Press, 2004.

Gloss, Gerry, Barbara Downes, and Andrew Boulton. *Freshwater Ecology: A Scientific Introduction.* Malden, MA: Blackwell Publishing, 2004.

Voshell, J. Reese, Jr. *A Guide to Common Freshwater Invertebrates of North America.* Blacksburg, VA: McDonald and Woodward Publishing Company, 2002.

Woodward, Susan L. *Biomes of Earth: Terrestrial, Aquatic, and Human Dominated.* Westport, CT: Greenwood Press, 2003.

Worldwatch Institute, ed. *Vital Signs 2003: The Trends That Are Shaping Our Future.* New York: W.W. Norton & Co., 2003.

PERIODICALS

Darack, Ed. "Death Valley Springs Alive." *Weatherwise.* 58. 4 July-August 2005: 42.

Marshall, Laurence A. "Sacred Sea: A Journey to Lake Baikal." *Natural History.* 116. 9 November 2007: 69.

Pollard, Simon D, and Robert R. Jackson. "Vampire Slayers of Lake Victoria: African Spiders get the Jump on Blood-filled Mosquitoes. (*Evarcha culicivora*)." *Natural History.* 116. 8 October 2007: 34.

Springer, Craig. "The Return of a Lake-dwelling Giant." *Endangered Species Bulletin.* 32. 1 Feburary 2007: 10.

ORGANIZATIONS

Canadian Lakes Loon Survey, PO Box 160, Port Rowan, ON, Canada N0E 1M0; Internet: http://www.bsc-eoc.org/cllsmain.html

Environmental Defense Fund, 257 Park Ave. South, New York, NY 10010, Phone: 212-505-2100; Fax: 212-505-2375; Internet: http://www.edf.org

Envirolink, P.O. Box 8102, Pittsburgh, PA 15217; Internet: http://www.envirolink.org

Environmental Protection Agency, 401 M Street, SW, Washington, DC 20460, Phone: 202-260-2090; Internet: http://www.epa.gov

Freshwater Foundation, 2500 Shadywood Rd., Navarre, MN 55331, Phone: 612-471-9773; Fax: 612-471-7685; Internet: http://www.envirolink.org/resource.html?itemid=601&catid=5

Friends of the Earth, 1717 Massachusetts Ave. NW, 300, Washington, DC 20036-2002, Phone: 877-843-8687; Fax: 202-783-0444; Internet: http://www.foe.org

Greenpeace USA, 702 H Street NW, Washington, DC 20001, Phone: 202-462-1177; Internet: http://www.greenpeace.org

International Joint Commission, 1250 23rd Street NW, Suite 100, Washington, DC 20440, Phone: 202-736-9024; Fax: 202-467-0746, Internet: http://www.ijc.org

Izaak Walton League of America, 707 Conservation Lane, Gathersburg, MD 20878, Phone: 301-548-0150; Internet: http://www.iwla.org/

North American Lake Management Society, PO Box 5443, Madison, WI 53705-0443, Phone: 608-233-2836; Fax: 608-233-3186, Internet: http://www.nalms.org

Project Wet, 1001 West Oak, Suite 210, Bozeman, MT 59717, Phone: 866-337-5486; Fax: 406-522-0394; Internet: http://projectwet.org

Sierra Club, 85 2nd Street, 2nd fl., San Francisco, CA 94105, Phone: 415-977-5500; Fax: 415-977-5799, Internet: http://www.sierraclub.org

World Wildlife Fund, 1250 24th Street NW, Washington, DC 20090-7180, Phone: 202-293-4800; Internet: http://www.wwf.org

WEB SITES

FAO Fisheries Department: http://www.fao.org/fi (accessed on September 5, 2007).

National Geographic Magazine: http://www.nationalgeographic.com (accessed on September 5, 2007).

National Park Service: http://www.nps.gov (accessed on September 5, 2007).

National Oceanic and Atmospheric Administration: http://www.noaa.gov (accessed on September 1, 2007).

Nature Conservancy: http://www.nature.org (accessed on September 5, 2007).

Scientific American Magazine: http://www.sciam.com (accessed on September 5, 2007).

UNESCO: http://www.unesco.org/ (accessed on September 5, 2007).

Ocean

The oceans are great interconnected bodies of salt water that cover 71 percent of Earth's surface, a total of 139,400,000 square miles (361,100,000 square kilometers). They contain 97 percent of all the water on Earth, a total volume of 329,000,000 cubic miles (1,327,000,000 cubic kilometers).

Oceanography, the science of the oceans, officially started in the 1870s when the British ship *Challenger* began its career of oceanic exploration. The findings of this expedition took twenty years to analyze and are published in fifty thick volumes. Another research vessel, the *Meteor,* equipped with electronic equipment, began exploration in 1925 and discovered great mountains and trenches on the sea floor. Only since the 1930s have people entered the deeper regions. The deepest spot, the Mariana Trench in the Pacific Ocean, was not reached until 1960. New information continues to be discovered all the time about how oceans work and what lives in them.

How the Oceans Were Formed

There are three main reasons why water covers so much of our planet. First, millions of years ago as Earth was forming, many active volcanoes released water vapor into the atmosphere in the form of steam. Second, Earth's gravity did not allow the water vapor to escape into space. It collected, along with other gases, to form clouds. Third, as Earth cooled, the moisture in those clouds condensed (turned from vapor to water), falling to Earth as rain. The rain filled the low areas in Earth's crust, and the cooler temperatures allowed much of this water to remain in liquid form. Over time, enough water accumulated to create a great ocean. If all of these factors had not been at work over 200 million years ago, our Earth might be dry, barren, and lifeless much like the moon.

WORDS TO KNOW

Bathypelagic zone: An oceanic zone based on depth that ranges from 3,300 to 13,000 feet (1,000 to 3,000 meters).

Coriolis Effect: An effect on wind and current direction caused by Earth's rotation.

Epigelagic zone: An oceanic zone based on depth that reaches down to 650 feet (200 meters).

Fast ice: Ice formed on the surface of the ocean between pack ice and land.

Hadal zone: An oceanic zone based on depth that reaches from 20,000 to 35,630 feet (6,000 to 10,860 meters).

Littoral zone: The area along the shoreline that is exposed to the air during low tide; also called intertidal zone.

Mesopelagic zone: An oceanic zone based on depth that ranges from 650 to 3,300 feet (200 to 1,000 meters).

Neap tides: High tides that are lower and low tides that are higher than normal when the Earth, sun, and moon form a right angle.

Neritic zone: That portion of the ocean that lies over the continental shelves.

Pack ice: A mass of large pieces of floating ice that have come together on an open ocean.

Spring tides: High tides that are higher and low tides that are lower than normal because the Earth, sun, and moon are in line with one another.

Thermocline: Area of the ocean's water column, beginning at about 1,000 feet (300 meters), in which the temperature changes very slowly.

Scientists named the original great ocean Panthalassa (pan-thah-LAHS-uh). More than 220 million years ago, only one large continent existed in this vast, primitive ocean. This land mass was named Pangaea (pan-GEE-uh). As time passed, Pangaea began to pull apart, and the ocean flowed into the spaces created between the land masses.

The breakup of Pangaea was caused by heat forces welling up from deep within Earth. Earthquakes split the ocean floor, creating fracture zones, or faults. Magma (molten rock) from below Earth's crust flowed into the fractures. As the magma cooled, it solidified, creating basins and ridges. After millions of years of repeated earthquakes and welling up of magma, the upper parts of Earth's crust were pushed farther apart. About 50 million years ago, the continents and the oceans took the basic shapes and positions of their current location.

The sea floor is still spreading at a rate of about 2 inches (5 centimeters) a year. This means the shapes of the continents and oceans are still changing.

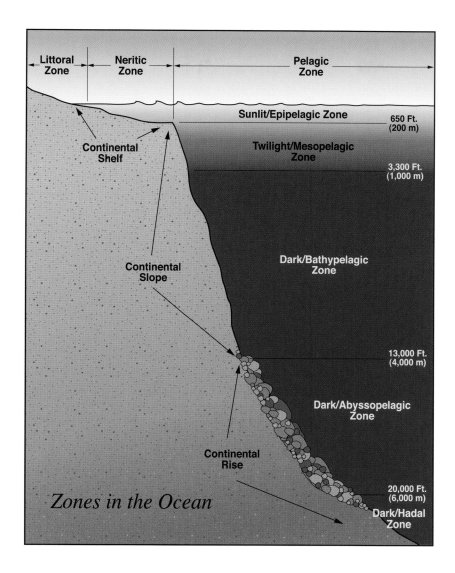

Zones in the Ocean

Major Oceans and Adjoining Regions

All the oceans considered together are called the World Ocean. The World Ocean is divided by the continents into three major oceans; the Atlantic, the Pacific, and the Indian. The Atlantic lies to the east of North and South America, and the Pacific lies to the west. The Indian Ocean lies to the south of India, Pakistan, and Iran. Some scientists consider the water

Goodbye, Baja

About 6 million years ago Baja (BAH-hah), California, split away from Mexico to form the Gulf of California. This splitting continues, and Baja is moving westward at a rate of about 2.5 inches (6 centimeters) a year.

surrounding the North Pole as a fourth ocean, the Arctic; most consider it part of the Atlantic.

The Pacific Ocean is the world's largest at 59,000,000 square miles (153,000,000 square kilometers). The next in size is the Atlantic at 32,000,000 square miles (83,000,000 square kilometers). The smallest of the three is the Indian Ocean at 26,000,000 square miles (67,000,000 square kilometers). The Atlantic is divided into North Atlantic and South Atlantic regions, and the Pacific into North Pacific and South Pacific.

The terms ocean and sea are often used interchangeably. However, an ocean is generally larger and deeper than a sea, and the physical features along its floor may be different. Seas are either contained within a larger ocean or connected to it by means of a channel. For example, the Sargasso, the Mediterranean, and the Caribbean Seas are all part of the Atlantic Ocean.

Even smaller bodies of water, called gulfs and bays, lie along the oceans' margins. A gulf is partly surrounded by land and it usually joins the ocean by means of a strait, which is a narrow, shallow channel. The Gulf of Mexico lies at the point where the Atlantic curves in between Mexico and Florida. It connects to the Atlantic by the Straits of Florida. A bay is partly enclosed by land, but it joins the ocean by means of a wide mouth or opening. The Bay of Bengal, which lies between India and Myanmar, is an example.

The Water Column

The water in an ocean, exclusive of the sea bed or other landforms, is referred to as the water column. The average depth of the World Ocean's water column is 12,175 feet (3,711 meters).

The term sea level refers to the average height of the sea when it is halfway between high and low tides and all wave motion is smoothed out. Sea level changes over time. Between 1930 and 1950, sea level along the east coast of the United States rose about 0.25 inches (1 centimeter) per year.

Ocean composition   Every element known on Earth can be found in the ocean. Ocean water is 3.5 percent dissolved salts by weight, including primarily chloride, sodium, sulfur (sulfate), magnesium, calcium, and

SEAS, GULFS, AND BAYS OF THE OCEANS

Atlantic Ocean	Indian Ocean	Pacific Ocean
Baltic Sea	Andaman Sea	Bering Sea
Barents Sea	Arabian Sea	Celebes Sea
Mediterranean Sea	Phillipine Sea	Coral Sea
Black Sea	Red Sea	East China Sea
Norwegian Sea	Sea of Okhotsk	Java Sea
Sargasso Sea	Sea of Japan	
North Sea	South China Sea	
Scotia Sea	Timor Sea	
Weddell Sea	Yellow Sea	
Greenland Sea	Gulf of Aden	
Gulf of Guinea	Gulf of California	
Gulf of St. Lawrence	Bay of Bengal	
Labrador Sea	Great Australian Bight (bay)	
Baffin Bay		
Hudson		

potassium. The measure of these salts determines the ocean's salinity (saltiness). One cubic mile (4.1 cubic kilometers) of sea water contains enough salt to cover all the continents with a layer 500 feet (153 meters) deep. These salts make sea water heavier than fresh water.

Most of the oceans' salts come from the weathering of rocks and the materials released by volcanoes. The level of saltiness has been increased by millions of years of evaporation and precipitation (rain, snow, and sleet) cycles. The water closest to the surface is usually less salty because of rainfall and fresh water flowing in from rivers. One of the saltiest bodies of water is the Red Sea, which receives little fresh water from rivers. One of the least salty is the Baltic Sea, which receives inflow from many rivers.

Water temperature   The temperature of the oceans varies. In general, temperature changes are greatest near the surface where the heat of the sun is absorbed. In the warmest regions, this occurs to depths of 500 to 1,000 feet (152 to 305 meters). Near the equator, the average surface temperature is about 77°F (25°C). The warmest surface water is found in

Taking the Salt from Sea Water

Water, water, everywhere, and all the boards did shrink;

Water, water, everywhere, nor any drop to drink.

The mariners (sailors) in Samuel Taylor Coleridge's (1772–1834) "Rime of the Ancient Mariner," were in trouble because they were out of fresh water. Sea water is not safe to drink because its high salt content causes illness and, eventually, death. In some areas with few fresh water sources, the salt can be removed from ocean water. This is process is called desalination.

There are several methods for desalination. In one method, the water is heated until it evaporates. As the water vapor condenses, it is collected as fresh water and the salts are left behind. In a second method, an electric current is passed through the water. The salts collect on strips of metal called electrodes. In a third method, the water is frozen and the salt crystals are separated from the ice crystals. The ice crystals are then melted back to water.

the Pacific Ocean. In February, off the coast of Australia, the surface temperature is about 81°F (27°C).

The average surface temperature of the ocean near the poles is about –28°F (–2°C). Sea water grows colder more slowly than fresh water because of its salt content. In the Arctic, the water column is permanently covered with ice. Pack ice (a mass of large pieces of floating ice that have come together) forms on open water. Fast ice is ice formed between pack ice and the land.

The range of annual temperatures on land can vary more than 100°F (47°C). In the ocean, temperatures vary only about 15°F (7°C) annually. Most of the ocean (about 95 percent) is so deep it is unaffected by the sun and seasonal changes. At a depth of about 1,650 feet (500 meters) begins an area called the thermocline, where temperatures change very slowly. Close to the ocean floor, the average water temperature is between 32° and 41°F (0° and 5°C) throughout the World Ocean.

Zones in the ocean   Different parts of the ocean have different features, and different kinds of creatures live in them. These different parts are called zones. Some zones are determined by the amount of light that reaches them. Others are based on depth and the life forms present. The different zones overlap and interact with one another.

Zones determined by light penetration   The surface waters of the ocean receive enough light to support photosynthesis in plants. Photosynthesis is the process by which plants use the energy from sunlight to change water and carbon dioxide (from the air) into the sugars and starches they use for food. These surface waters are called the sunlit zone, which reaches down as far as 660 feet (200 meters) below the surface.

The next level is the twilight zone, which ranges from 660 to 3,000 feet (200 to 914 meters). Only blue light can filter down to this level, where it is too dark for plant life, but where at least 850 species of fish make their home. These twilight dwellers have large eyes and often travel to the sunlit zone at night to feed.

In the deepest region of the oceans, from 3,000 to 36,163 feet (914 to 11,022 meters), there is no light. This is called the dark zone. It is not known if any fish that live here travel upward at night, but their eyes are usually tiny and many are blind.

Zones determined by depth Based on depth, the zone closest to the surface of the sea is called the epipelagic zone. It reaches down to 660 feet (200 meters) and corresponds to the sunlit zone. The next zone is the mesopelagic zone, and it corresponds to the twilight zone. It ranges from 660 to 3,000 feet (200 to 914 meters). The zones of darkness follow. First is the bathypelagic zone, which covers from 3,000 to 9,843 feet (914 to 3,000 meters). Then comes the abyssopelagic zone, ranging from 9,843 to 19,685 feet (3,000 to 6,000 meters). At the very bottom is the hadal zone. It reaches from 19,685 feet (6,000 meters) to the very bottom of the Mariana Trench, the deepest spot on Earth at 35,840 feet (10,924 meters).

Zones determined by sea life The entire water column of the ocean is a vast ecosystem (an environment in which all organisms living within are dependent on other living and nonliving organisms for survival and continued growth) called the pelagic zone. Most life forms that live in it are concentrated near the surface where light is available. The neritic zone is the portion of the ocean that lies over the continental shelves (extensions of the continent that taper gently into the sea). The littoral zone refers to the area along the shoreline that is exposed to the air during part of the day as tides flow in and out.

Ocean circulation
The oceans are constantly, restlessly moving. This movement takes the form of tides, waves, surface currents, vertical currents, and eddies and rings.

Tides Tides are rhythmic movements of the oceans that cause a change in the surface level of the water, noticeable particularly along the shoreline. When the water level rises, it is called high tide. When it recedes (drops), it is called low tide. Some tides, such as those in the Mediterranean Sea, are barely measurable. In the Bay of Fundy in Nova

Frogfish are dark-zone fish.
IMAGE COPYRIGHT TILL VON AU, 2007. USED UNDER LICENSE FROM SHUTTERSTOCK.COM.

Melting Ice

The great icecap over the North Pole has shrunk in size during different prehistoric periods, adding its water to the oceans. This added water has raised sea level as much as 660 feet (200 meters). During cooler periods the icecap has always been restored to an even larger size.

Scotia, the difference between high and low tide may be as much as 52 feet (16 meters). High and low tides occur in a particular place at least once during each period of 24 hours and 51 minutes.

Tides are caused by a combination of the gravitational pull of the sun and moon, and Earth's rotation. The gravitational pull from the sun or moon causes the water to bulge outward. At the same time, the centrifugal force created by Earth's rotation causes another bulge to occur on the opposite side of Earth. These areas experience high tides. Water is pulled from the areas in between, and those areas experience low tides.

When the Earth, sun, and moon are lined up, the gravitational pull is stronger. Then, high tides are higher and low tides are lower than normal. These are called spring tides. When the Earth, sun, and moon form a right angle, the gravitational pull is weaker. This causes high tides to be lower and low tides to be higher than normal. These are called neap tides.

Wind-driven surface waves   Waves are rhythmic rising and falling movements of the water. Although waves make the water appear as if it is moving forward, forward movement is actually very small. Most surface waves are caused by wind. Their size is due to the speed of the wind, the length of time it has been blowing, and the distance over which it has traveled. As these influences grow stronger, the waves grow larger, and storm waves can produce waves over 100 feet (31 meters) high. When the ocean's surface can absorb no more energy, instead of growing in size, the waves collapse. Hurricanes, with wind speeds of 106 miles per hour (170 kilometers per hour), rarely raise waves higher than 43 feet (13 meters).

Once set in motion, waves can move for long distances. Over time, they become more regular in appearance and direction, forming a swell. By studying the movement of the swell, experienced ocean travelers can determine where a distant storm has occurred.

In shallow areas, such as along a shoreline, the bottom of a wave is slowed down by friction as it moves against the sea floor. The top of the wave is not slowed down by friction and moves faster than the bottom. When the top finally gets ahead of the bottom, the wave tumbles over on itself and collapses causing a breaker and sending a mass of swirling, bubbling foam tumbling onto the shore.

Tsunamis A type of surface wave called a tsunami (soo-NAH-mee) is caused by undersea earthquakes. When the ocean floor moves during the quake, its vibrations create a powerful wave that travels to the surface. The tsunami is barely noticeable in mid-ocean, but as it approaches bays, channels, or sloping shorelines, its power is concentrated. Suddenly its height increases, sometimes forming a towering crest that can reach a height of 200 feet (61 meters) as it crashes onto the land. A tsunami that struck eleven countries bordering the Indian Ocean in 2004 was 50 feet (15 meters) high and killed over 130,000 people.

Most tsunamis do not create walls of water but appear as sudden upwellings (rising of the water level). They are seldom just one wave. A dozen or more that vary in strength often travel in succession. Tsunamis can move as fast as an airplane—440 miles (700 kilometers) per hour—and can travel thousands of miles from their source before sweeping onto the land. As they recede back to sea, they make a loud sucking noise.

When tsunamis strike inhabited areas, they can destroy entire towns and kill many people. Some people drown as the wave washes inland; others are pulled out to sea when the tsunami recedes. Hawaii is very vulnerable to tsunamis because of its position in a Pacific region known for frequent volcanic activity and earthquakes. The Pacific Ocean experiences, on average, two life-threatening tsunamis each year. They are monitored by the Pacific Tsunamis Warning Center (PTWC).

Surface currents Currents are the flow of water in a certain direction. They can be both large and strong. For example, the Gulf Stream, a current that lies off the eastern coast of the United States, and the Kuroshio (koo-ROH-shee-oh) near Japan, travel at 2.5 to 4.5 miles per hour (4 to 7 kilometers per hour). Usually, surface currents do not extend deeper than a few hundred yards (183 meters). The Florida Current and the Gulf Stream extend to depths of 6,560 feet (2,000

Storm at Sea

The large and usually violent tropical cyclones that form over oceans are called hurricanes in the Atlantic and eastern Pacific, and typhoons in the western Pacific. Their wind speeds are at least 75 miles (120 kilometers) per hour and may reach 180 miles (300 kilometers) per hour. A single storm may cover an area up to 2,000 miles (3,200 kilometers) in diameter and release up to 10 inches (25 centimeters) of rain a day.

Hurricanes and typhoons require very moist air to supply their energy, and only very warm air contains enough moisture. As a result, they only form over water at temperatures of at least 80°F (27°C). As they move over cooler regions, their power diminishes and they break up.

The forceful, rotating winds created by tropical storms cause much damage, especially when they reach land, as do the waves that batter the shoreline. Many people lose their lives each year in tropical storms. After 1944, airplanes were used to help spot and keep track of these storms. Weather satellites in orbit around Earth now monitor their growth and movement.

The Sea State Scale

The heights of waves are measured on the sea state scale in number values from zero to nine. A sea state of zero means calm, smooth water. A sea state of nine means waves over 45 feet (14 meters) high.

meters). They are caused by the wind, the rotation of Earth, and the position of continental landmasses. They contain about 10 percent of the World Ocean water.

Effects of wind   Winds directly affect only the upper zone of the ocean down to about 660 feet (200 meters). The currents created by the wind may reach depths of more than 3,000 feet (914 meters). Wind-driven currents move horizontally (parallel to Earth's surface).

Winds over the oceans tend to follow a regular pattern, generally occurring in the same place and blowing in the same direction. At the equator are the doldrums, very light winds that create little water movement. Both north and south of the equator to about 30° latitude are the trade

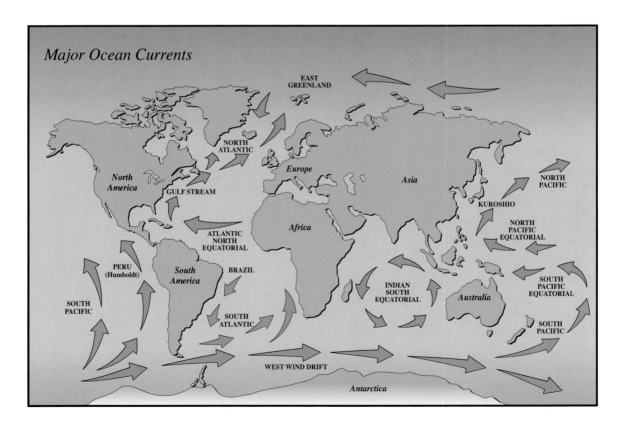

Major Ocean Currents

winds, which blow primarily east. (Latitude is a distance north or south of Earth's equator, measured in degrees.) At about 60° latitude are the westerlies. Near the poles, the polar easterlies occur. These three wind patterns create three basic systems of ocean currents: the equatorial system, the subtropical gyre (JYR), and the subpolar gyre. (A gyre is a circular or spiral motion.)

The Coriolis effect   The rotation of Earth influences the patterns of the wind and the ocean's currents. This is called the Coriolis (kohr-ee-OH-lus) Effect. In the Northern Hemisphere, the Coriolis Effect causes air masses moving south to veer westward. Just the opposite happens in the Southern Hemisphere. Air masses moving north veer eastward. Currents north of the equator move in a clockwise direction. Currents south of the equator move counterclockwise. The Coriolis Effect is greater near the poles and causes the currents on the western sides of ocean basins to be stronger than those on the eastern sides.

Currents that develop from Coriolis forces are usually permanent. They include the West Wind Drift, the North Equatorial Current, and the South Equatorial Current.

Effect of landmasses   Currents are affected by the presence of obstacles such as the continents and large islands. In the North Pacific, for example, currents moving west are deflected northward by Asia and southward by Australia. The same currents then move east until North and South America send them back toward the equator.

Vertical currents   Upward and downward movements occur in the ocean. These vertical currents are primarily the result of differences in water temperature and salinity.

At the North and South Poles, the surface water often freezes, due to its lower salt content. As a result, the salts become more concentrated in the water below the surface, causing this water to remain unfrozen and become heavier. This heavy, cold water then sinks and travels along the ocean floor toward the equator. At the same time, near the equator, the sun warms the surface water, which travels toward the poles. As this cycle repeats itself, the water is continually circulating. Warm surface water flows toward the poles, cools, sinks, and flows back toward the equator. Vertical currents help limit the depth of horizontal currents.

Tides of Fire

In 1964, an earthquake struck Alaska, causing great damage. A pipeline carrying oil was broken and the oil caught fire. A tsunami three stories high followed the quake, surging inland. Because oil floats on water, the burning oil was carried overland in a tide of fire. The fires reached the railroad yards, where the iron tracks soon glowed red from the heat. More tsunamis followed and flowed over the tracks. The sudden cooling made the tracks rise up and curl like giant snakes

Ben Franklin Speeds the Mail

In 1770, Benjamin Franklin (1706–1790) was postmaster general for the American colonies. At that time, the colonies depended upon sailing ships to carry the mail back and forth between England and America. A constant complaint from postal customers was that mail going east to England always arrived weeks sooner than mail moving west to the colonies. Puzzled, Franklin decided to investigate.

Franklin's cousin, Timothy Folger, was a ship's captain. He told Franklin about the presence of the Gulf Stream. He explained that when they traveled to England, the ships moved with the current, which added to their speed. When they returned to America, they had to fight the current, which slowed them down.

Luckily, the Gulf Stream was only about 62 miles (100 kilometers) wide. Franklin believed it could be avoided and mapped a new route. He named the current the Gulf Stream because it flowed out of the Gulf of Mexico.

In some coastal areas, strong wind-driven currents carry warm surface water away. An upwelling of cold water from the deep ocean occurs to fill the space. This is more common along the western sides of the continents. These upwellings bring many nutrients from the ocean floor to the surface.

Eddies or rings   Eddies or rings are whirl-pools that move in a circular motion against the flow of a main surface current. Eddies are probably caused when the speed and intensity of a current increases to such a point that the current becomes unstable. The eddy allows the excess speed and intensity to be distributed into the surrounding water.

Effect of the water column on climate and atmosphere   The World Ocean is responsible for much of the precipitation (moisture) that falls on land. Ocean water evaporates in the heat of the sun, forms clouds, and falls elsewhere in the form of rain, sleet, or snow.

The ocean absorbs and retains some of the sun's heat. In the winter, this heat is released into the atmosphere, helping to keep winter temperatures warmer inland. In summer, when the water temperature is cooler than the air temperature, winds off the ocean help cool coastal areas.

Pools of warmer or cooler water within the ocean influence storms in the atmosphere. The *El Niño* flow of warmer water, which affects the storm systems in North and South America, is an example. This warm current causes changes in winds and air pressure, bringing severe storms and droughts (very dry periods). How *El Niño's* effect comes about is still not clearly understood.

The oceans help regulate the levels of different gases in the atmosphere, such as oxygen and carbon dioxide. About one-half of the carbon dioxide that is produced by burning forest fires and other causes is captured by the oceans. Too much carbon dioxide in the atmosphere contributes to warmer global temperatures. The ocean's presence helps moderate such undesirable changes.

Geography of the Ocean Floor

The ocean floor is that area of Earth's crust covered by ocean water. It is divided into the continental margins and the deep-sea basins.

Continental margins   The continental margin is the part of the seafloor at the edges of the continents and major islands where, just beyond the shoreline, it tapers gently into the deep sea. The continental margin includes the continental shelf, the continental slope, and the continental rise.

Continental shelf   The continental shelf is that part of the margin that begins at the shoreline. It is flat and its width varies. Off the coast of Texas the shelf is 125 miles (200 kilometers) wide. The continental shelf is usually less than 660 feet (200 meters) below sea level. It receives much rich sediment (soil and other particles) from rivers that flow to the sea and, being in the sunlit zone, it supports many forms of ocean life.

Continental slope   At the end of the continental shelf there is a steep drop. This is the continental slope, which descends to depths of 10,000 to 13,000 feet (3,000 to 4,000 meters) and ranges from 12 to 60 miles (20 to 100 kilometers) in width. In many places, the slope is cut by deep underwater canyons that may have been formed by prehistoric rivers.

Continental rise   Beyond the continental slope is the continental rise, where sediments drifting down from the continental shelf collect. These deposits may extend as far as 600 miles (914 kilometers) into the ocean where the deep-sea basin begins.

Deep-sea basins   The deep-sea basins begin at the edge of the continental rise. These vast, deep basins in the ocean floor contain underwater mountain ranges (ridges), volcanoes, deep trenches, and wide plains.

The mid-ocean ridge   A long chain of submerged (underwater) mountains called the mid-ocean ridge runs through the World Ocean. This ridge was formed when rifts (cracks) were created in the sea floor by earthquakes and volcanic action. Hot lava seeped or poured out of these cracks, spreading apart the sea floor. As the seafloor spread, it buckled, forcing Earth's crust upward and forming a chain of mountains about 40,000 miles (65,000 kilometers) long. This ridge was discovered at different locations by different scientists who did not realize it was one chain and who gave it different names, such as mid-Atlantic ridge or West Chile Rise.

LANDFORMS OF THE OCEAN'S BASINS

Atlantic Ocean Basin	Indian Ocean Basin	Pacific Ocean Basin
Falklands Escarpment	Central Indian Ridge	Aleutian Trench
Mid-Atlantic Ridge	Java Trench	East Pacific Rise
Puerto Rico Trench	Ninetyeast Ridge	Emperor Seamount Chain
Sandwich Trench		Hawaiian Ridge
Scotia Ridge		Japan Trench
Walvis Ridge		Kermadec-Tonga Trench
		Kuril Trench
		Mariana Trench
		Peru-Chili Trench
		Philippine Trench
		West Chile Trench

The mid-ocean ridge usually rises 0.6 to 1.8 miles (1 to 3 kilometers) above the surrounding sea floor. Those peaks that break the surface of the water form volcanic islands. Iceland is an example. The highest submerged mountain is 29,520 feet (8,997 meters) and is found between Samoa and New Zealand.

Deep valleys often cut into the mid-ocean ridge, and are frequently the site of volcanic and earthquake activity.

Seamounts   Some isolated volcanoes that do not reach the surface of the ocean, called seamounts, form alongside the mid-ocean ridge but are separate from it. Seamounts have steep slopes and may be as high as 13,000 feet (5,000 meters).

Smokers   On the deep ocean floor in highly volcanic zones, jets of hot water (up to 716°F [380°C]), called smokers, have been discovered. These water jets are composed of ordinary seawater that enters clefts (splits) in hot volcanic rock where it heats up, expands, and escapes. In its passage through the rock, the hot water absorbs large quantities of dissolved minerals that make it look cloudy and smokelike as it reenters the surrounding ocean. Sometimes the dissolved minerals are deposited around the vents (openings) through which it is expelled, forming hollow black chimneys as high as 33 feet (10 meters).

Oceanic islands rise from the ocean floor, often as the result of volcanic activity. IMAGE COPYRIGHT SPECTA, 2007. USED UNDER LICENSE FROM SHUTTERSTOCK.COM.

Trenches   As the sea floor spreads, it meets the edges of the continents, which resist its movement. This results in an area of extreme pressure called a subduction zone. In the subduction zone, this enormous pressure forces the expanding sea floor down and under the continental margin, often causing a deep, V-shaped trench to form. The greatest depths in the oceans are found in these trenches, and the deepest trenches are located in the Pacific. The Mariana Trench is the deepest at 35,840 feet (10,924 meters).

Ocean trenches are much deeper than any valley on the continents. The Grand Canyon in Arizona, for example, is about 1 mile (1.6 kilometers) deep. The Mariana Trench in the Pacific is about 6.8 miles (11 kilometers) deep. If Mt. Everest, the tallest mountain on Earth, could be put into the Mariana Trench, more than 6,800 feet (2,073 meters) of ocean would still cover it.

Islands and atolls   Continental islands were once part of a nearby continent. As the continent drifted, the island broke away. These islands may include hills and mountains similar to those on the continent.

Oceanic islands are those that rise from the deep-sea floor. Volcanoes, some of which are still active, formed most of the islands. At first, volcanic islands lack any form of life except possibly bacteria. Seeds and plant spores (single cells that grow into a new organism) arrive on the

The Newborn Island, Surtsey

In 1963, a baby island was born off the coast of Iceland. It has been named Surtsey. The product of an underwater volcano, Surtsey was formed of cooled magma. A similar island may be forming on the seabed southeast of Hawaii.

wind or are carried there by ocean waves or visiting animals. Gradually, an ecosystem tailored to each particular island is formed.

Atolls are ring-shaped coral reefs that have formed around a lagoon. A coral reef is created by small, soft, jellylike animals called corals. Corals attach themselves to hard surfaces and build a shell-like external skeleton. Many corals live together in colonies. Young corals build their skeletons next to or on top of older skeletons. Over hundreds, thousands, or millions of years, a wall, or reef, of these skeletons is formed. The mountain peak in the center gradually sinks or is washed away and only the reef remains.

Abyssal plains   Abyssal plains are the vast flat areas where light does not penetrate. They make up the largest portion of the ocean floor. The abyssal plain is deepest in the Pacific Ocean at 20,000 feet (6,098 meters) and shallowest in the Atlantic Ocean at 10,000 feet (3,048 meters).

Where the ocean floor remains uncovered by sediment, a hard rock called basalt is visible. Elsewhere, the floor is covered by sediments that have drifted down from the continental margins, from volcanic activity, from dead marine life, from coal-burning ships, and even from dust carried by the wind and deposited on the water's surface. These sediments build up until they create a flat surface.

Sediments are of two types. Those formed from the waste products and dead tissues of plants and animals are called ooze. Usually, ooze is found only in temperate (having moderate temperatures) and tropical (hot and humid) regions where it may become hundreds of yards (meters) thick. Other sediments are usually red clay. About 1 inch (2.5 centimeters) of red clay is deposited on the ocean floor every 2,500 years. In parts of the ocean floor, the sediment layer is more than 3,000 feet (7,000 meters) thick.

Plant Life

Plants that live in the sea are surrounded by water at all times. For this reason, they have no need to develop the special tissues and organs for conserving water that are needed by plants on land. Seaweeds, for example, use their rootlike structures called holdfasts to anchor them in one spot. They do not use them to draw water from the soil.

The Web of Life: Alien Algae

In 1984, an alien was discovered in the Mediterranean Sea. Although probably not from outer space, its origin remains unknown and it is gradually taking over. It is a species of green algae that usually grows only in aquariums. Somehow it found its way into the sea where it grows aggressively with no natural enemies. It is smothering anything from the beach to a depth of about 150 feet (46 meters). It has no scientific name, but some researchers believe it is derived from *Caulerpa taxifolia*. Others are in favor of calling it *C. Godzilla* for obvious reasons.

This species reproduces by fragmentation; small pieces that break off grow into new plants. Any attempt to dig it up or mow it down fails, because millions of potentially new plants are created in the process. Pesticides are too toxic and dangerous for wildlife. The only hope for saving the Mediterranean from the algae seems to be the use of bio-controls (natural means of controlling pests). Scientists are searching for an animal, such as a snail or slug that will eat the algae. Bio-controls can cause other problems. Sometimes the new animal turns up its nose at the pest plant and eats other, more desirable plants. Then, instead of one problem, there are two. Until a remedy is found, many countries, including the United States, are prohibiting the importation of this type of algae for use in aquariums.

The water offers support to ocean plants. Giant underwater plants do not require tough woody stems like land plants to remain upright; the water holds them up. Instead, sea plants have soft and flexible stems, allowing them to move with the current without breaking.

Many species of plants live in the oceans; however, all species are not found in all oceans. Species of plants found in the Indian Ocean, for example, are seldom found in the Atlantic. Most plants live along the continental margins and shoreline, not in mid-ocean.

Ocean plants can be divided into two main groups: plantlike algae (AL-jee) and green plants.

Algae   It is generally recognized that algae do not fit neatly into the plant category, but they have some characteristics similar to true plants.

Algae are found extensively in the oceans. Most algae have the ability to make their own food by means of photosynthesis (foh-toh-SIN-thih-sihs). Others absorb nutrients from their surroundings.

The types of algae commonly called seaweed resemble green plants, but they have no true leaves, stems, or roots. Some are so tiny they cannot be seen with the naked eye. Others are massive and live in vast underwater forests. Many algae, such as kelp, have soft, jellylike cell surfaces. Other types, such as diatoms, form shells, scales, or stony coverings.

Seaweed provides nutrition and also homes for many animals living in the ocean.

IMAGE COPYRIGHT GRAEME KNOX, 2007. USED UNDER LICENSE FROM SHUTTERSTOCK.COM.

Some algae float freely in the water, allowing it to carry them from place to place. These are called plankton (a Greek word meaning "wanderers"). Kelp anchors itself to the sea floor, and belongs to the benthic species. (Benthos in Greek means "seafloor.")

Common algae Phytoplankton are tiny plants that usually cannot be seen without the aid of a microscope. They float on the water's surface, always within the sunlit zone, and are found near the coast and in mid-ocean. They are responsible for about 90 percent of the photosynthesis carried out in the oceans, helping to supply Earth's atmosphere with oxygen as a result.

Two forms of phytoplankton are the most common, diatoms and dinoflagellates (dee-noh-FLAJ-uh-lates). Diatoms have simple, geometric shapes and hard, glasslike cell walls. They live in colder regions and even within Arctic ice. Dinoflagellates have two whiplike attachments that make a swirling motion to help them move through the water. They live in tropical zones (regions around the equator).

Growing season Both land and sea plants contain chlorophyll, the green pigment in which they turn energy from the sun into food. As long as light is available, ocean plants can grow, even in the Arctic. In some algae, the green of chlorophyll is masked by orange-colored pigments, giving them a red or brown color.

Food Most algae grow in the sunlit zone where light is available for photosynthesis. Algae require other nutrients that must be found in the water, such as nitrogen, phosphorus, and silicon. In certain regions, upwelling of deep ocean waters during different seasons brings more of these nutrients to the surface. This results in greater numbers of algae during those times. These increases are called algal blooms. When the upwelling ceases, their numbers decrease.

Nitrogen and phosphorus are in the shortest supply in most oceans, which limits plant growth. When these elements are added to a body of water by sewage or by runoffs from farmland, there is a sudden burst of algae growth in these regions.

Reproduction Algae may reproduce in one of three ways. Some split into two or more parts, each part becoming a new, separate organism. Others produce spores. A few reproduce sexually; cells from two different plants unite to create a new plant.

Green plants Green plants do not grow in mid-ocean. Instead, various types of seagrasses live in protected areas along the continental shelf where their roots can find soil and nutrients. Some marine animals use them for food and for hiding places.

Animal Life

The oceans swarm with life and are the largest animal habitats on Earth. More life forms are found in the oceans than in tropical rain forests. Most species live along the continental margins. Some of those enter mid-ocean regions when they migrate to different areas in search of food or for breeding purposes. The numbers that live in mid-ocean permanently are fewer. The Gulf Stream, for example, supports so few species it is almost like a desert.

Living organisms must maintain a balance of body fluids and salt levels in their blood. Animals that live in the sea have developed ways to cope with its high salinity. Naked animals or those with thin skins or shells usually have high levels of salt in their blood and do not need to expel any excess. Others, such as most fish, have special organs that remove the extra salt from their bodies and release it into the water.

The water offers support to marine (ocean) animals as it does to plants. For example, giant whales could not exist on land because of their body weight. The effect of gravity would be too great. But ocean water allows them to float and swim effortlessly. Many marine animals have special chambers in their bodies that allow them to adjust their buoyancy (BOY-un-see; ability to float) so that they can change their depth. Some, such as seals, have fins to make swimming easier. Others, such as octopi, forcefully eject water in a kind of jet stream to help them move.

Distribution of animals in the oceans depends primarily on the food supply. Tropical seas, which are rich in nutrients, have many more life

The Colors of Kelp

Kelp must absorb sunlight in order for photosynthesis to take place. For this reason, kelp are found in the sunlit zone in fairly shallow water. Different species are found at different depths, and these species vary in color. This color difference is due to the filtering effect of the water. As sunlight penetrates deeper, different colors in it are filtered out. Near the surface, kelp is green. Farther down, the brown varieties grow. Red and blue species are found where there is the least amount of light.

Small plants thrive in the ocean, providing food and protection for fish and small invertebrates. IMAGE COPYRIGHT STEPHEN KRAUSSE, 2007. USED UNDER LICENSE FROM SHUTTERSTOCK.COM.

Tidepools form at the edge of the ocean. During low tide, they are small ponds, separated from the ocean, but when they tide comes in, they are covered by the ocean again. IMAGE COPYRIGHT LEV RADIN, 2007. USED UNDER LICENSE FROM SHUTTERSTOCK.COM.

forms than polar seas. The lack of sudden temperature changes in the oceans makes the environment comfortable for most marine animals.

Marine animals may be classified as microorganisms, vertebrates, and invertebrates. They are also classified according to their range of movement. Plankton drift in the currents and include both microorganisms and larger animals such as jellyfish. Many plankton move upward and downward in the water column by regulating the amount of gas, oil, or salt within their bodies. (Production of gas or oil and removal of salt causes the organism to rise; the reverse causes it to sink.) Many larger animals spend part of their young lives as plankton. Crabs, which float with the current in their larval (immature) form, are an example. Larger animals that swim without the help of currents, such as fish and whales, are called nekton. Benthos are animals that live on the ocean floor. These include snails, clams, and bottom-dwelling fish.

Microorganisms Microorganisms cannot be seen with the human eye without the aid of a microscope. Most microorganisms are zooplankton (tiny animals that drift with the current). They include foraminiferans, radiolarians, acantharians, and ciliates, as well as the larvae or hatchlings of animals that will grow much larger in their adult form.

Bacteria Until 1998, it was believed that bacteria appeared in the ocean only where other organisms were decomposing (decaying). Scientists now agree that bacteria are found throughout the ocean and make up much of the dissolved matter in the mid-ocean water column. As such, they provide food for microscopic animals. Bacteria help decompose the dead bodies of larger organisms, and for that reason their numbers increase near the coastlines where the larger organisms are found.

Bacteria are found in the sediment on the dark ocean bottom and around the waterjet smokers created by deep-sea volcanic vents. These deep-sea bacteria obtain food and oxygen by means of chemosynthesis (kee-moh-SIHN-thuh-sihs), a process in which the organism creates food using chemical nutrients as the energy source instead of sunlight. The bacteria live in cooperation with animals unique to this region, providing them with important nutrients.

Food Some zooplankton eat phytoplankton and are preyed upon by other carnivorous zooplankton, such as arrow worms. Krill are shrimplike zooplankton that play an important part in the food web of the ocean. They feed upon plant-eating zooplankton and are eaten in turn by larger fish and mammals. The largest mammals of all, the great whales, consume tons of krill each day.

Invertebrates Animals without backbones are called invertebrates. Many species are found in the oceans. Crustaceans, invertebrates with hard outer shells such as lobsters and clams, are probably the most numerous and diverse group.

Common invertebrates Invertebrates found in the oceans range from planktonic jellyfish to benthic worms and crabs.

Most crustaceans swim freely in the upper waters. Some, like the barnacle, attach themselves permanently to solid surfaces. Clams typically live in shallow water. At least one species has been found on the deep ocean floor near smokers where they have adapted to the warm, sulfur-rich water.

Squid live along coastlines and in mid-ocean, in surface and deep waters. They can exist as plankton, drifting along with the current, or use

The Red Tide

Some species of dinoflagellates are poisonous. When enough of them are present, they color the water red, creating what is called red tide. Their toxic secretions kill many fish, and even humans are not immune. When the poison is released into the air by waves breaking on shore, people can develop irritation in their mouth and lungs.

Ocean waters allow octopus and other sea creatures to swim effortlessly. IMAGE COPYRIGHT LAVIGNE HERVE, 2007. USED UNDER LICENSE FROM SHUTTERSTOCK.COM.

Sea anemone look like plants but they are actually invertebrates.
IMAGE COPYRIGHT VIRGINIA GOSSMAN, 2007. USED UNDER LICENSE FROM SHUTTERSTOCK.COM.

a form of jet propulsion by squirting water in a forceful stream from a tube near their heads.

Giant tubeworms grow up to 8 feet (2.4 meters) long and live in clusters around deep-sea smokers. These worms have neither a mouth nor a digestive system. They survive because of bacteria that live around them and produce the nutrients they need by chemosynthesis.

Food   Invertebrates may eat phytoplankton, zooplankton, or both. Some eat plants or larger animals. The sea anemone uses its stinging tentacles to paralyze shrimp and small fish that swim past. Other invertebrates, including lobsters and crayfish, roam along the ocean floor to feed on dead organisms. A species of hadal zone spider has no eyes and feeds by sticking a long tube into worms and other soft creatures and sucking out body juices.

Reproduction   Marine invertebrates reproduce in one of three ways:

- Eggs are laid and fertilized externally. A parent watches over the young in the early stages, and offspring number in the hundreds.

- Fertilization is internal. A parent cares for the young in the early stages, and offspring number in the thousands.

- Fertilization is external. The young are not cared for, and offspring number in the millions. Survival depends upon the absence of predators and the direction of currents.

Reptiles   Reptiles are vertebrates (animals having backbones). The body temperatures of reptiles changes with the temperature of their surroundings. In cold temperatures, they become sluggish, so they do not have to waste energy keeping their body temperatures up as do most mammals and birds.

The two types of reptiles found in the oceans are sea turtles and sea snakes. Sea turtles can be distinguished from land turtles by their paddle-shaped limbs called flippers, which enable them to swim. The largest marine turtle, the Atlantic leatherback, may weigh as much as 2.2 tons (2,020 kilograms) and measure 9 feet (3 meters) from the tip of one flipper to another. Sea turtles have glands around their eyes that remove excess salt from their bodies. This process makes them appear to cry.

At least thirty-two species of snakes live in tropical oceans, and all of these are found in regions around Australia and New Guinea. They look much like land snakes, having long bodies, some of which attain 9 feet (3 meters) in length. Unlike land snakes, they have salt glands that help them maintain a body fluid balance. Their tails are paddle shaped, which helps them move through the water, and some species are able to close their nostrils, which enables them to dive and remain submerged for hours. Sometimes, colonies of thousands can be seen floating on the surface of the water, basking in the sun.

Sea turtles are found in all of the world's oceans except the Arctic Ocean. IMAGE COPYRIGHT STEPHEN STRATHDEE, 2007. USED UNDER LICENSE FROM SHUTTERSTOCK.COM.

Food   Reptiles can survive for a long time without eating. When food is obtained they often eat large amounts. Turtles are omnivorous (eating both plants and animals). They eat soft plant foods as well as small invertebrates such as snails and worms. Turtles have no teeth. Instead, the sharp, horny edges of their jaws are used to shred food enough to swallow it.

Sea snakes are carnivorous (meat-eaters), feeding primarily on fish, including eels, that they find along the ocean floor in rocky crevices. They first bite their prey, injecting it with poison so it cannot escape.

Common reptiles   Green sea turtles are typical of oceanic reptiles. They are migrators, traveling as far as 1,250 miles (2,000 kilometers) to return to a particular breeding area where they lay their eggs. Analysis of their brain structure suggests that they have a keen sense of smell and a built-in sense of direction that guide them to their particular breeding grounds. One green turtle was observed to travel 300 miles (480 kilometers) in ten days, which indicates it was moving quite quickly. Green sea turtles and the other six species of sea turtles are endangered.

Sea snakes are all venomous. The most dangerous is the beaded sea snake since just three drops of its venom can kill nine people. Very few human deaths occur from sea snake bites because the creatures are afraid of humans.

Reproduction   Both snakes and turtles lay eggs, and some types of snakes bear live young. Turtles lay their eggs in a hole on a sandy shore, which they then cover. After the nest is finished, the female abandons it and seems to take no interest in the offspring. Six weeks later, the eggs

The Sneaky Shark

The cookie-cutter shark, a dwarf species named because it bites a round, cookie-shaped chunk of flesh out of bigger animals, uses a sneaky method of attracting prey, The cookie-cutter's underside glows, except for a large spot beneath its jaw. The ability to glow helps it blend in with the light filtering down from above. To another predator, the dark spot beneath its jaw may appear to be a smaller, innocent-looking fish. When a tuna or swordfish comes along, it spies the dark spot and thinks lunch is at hand. But when the big fish rushes upward for a bite, the cookie-cutter shark is the one that gets the meal.

hatch and the young turtles run for the water and disappear into the ocean.

Some species of sea snakes come to shore during breeding season and lay their eggs on land. The remainder bear live young at sea.

Fish   The oceans are home to about 14,700 kinds of fish, about 60 percent of all fish species on Earth. Fish are primarily cold-blooded (temperature varies with the surroundings) vertebrates having gills and fins. The gills are used to draw in water from which oxygen is extracted. The fins are the equivalent to arms and legs and are used to help propel the fish through the water.

Mackerel sharks and tuna are examples of warm-bodied fish. They have a special system of blood circulation that allows them to keep their body temperature as much as 21°F (12°C) above the water temperature. This extra warmth gives them more muscle power and thus more speed.

Fish vary in size from the tiny goby of the Philippines that measures less than 0.5 inch (1 centimeter) in length to the giant whaleshark that can grow as long as 60 feet (18 meters). Most fish have long, streamlined bodies built to move swiftly through the water, but shapes vary greatly. Manta rays, for example, are flat and round. Seahorses are narrow and swim with their bodies in a vertical (upright) position.

Most fish species live in the sunlit zone. Approximately 855 species live in the twilight zone, including the hatchetfish, the dragonfish, the lanternfish, the lancet fish, and the viperfish. These fish often have large, upward-angled eyes designed to see in the murky depths. They may have rows of light organs on their undersides, which help them blend in with the more well-lit waters above them or help them attract mates. The light from these organs is produced by either bacteria or chemical reactions they produce themselves. Some species have stretchy bodies, huge jaws, and fanglike teeth that enable them to take in and digest prey as large or even larger than themselves. These features may help them survive in a region where food is scarce.

Fish living in the dark zone include the gulper eel, the anglerfish, the whalefish, and the tripod fish. Most are white and small and either have no eyes or have tiny eyes that are blind. They, too, may have stretchy bodies and huge mouths. Few have light organs, perhaps because these would not be very useful around other blind fish. Many twilight and dark-zone fish have sensory hairs on their bodies that detect motion or changes in water pressure. Some can also detect odors.

Common fish   Typical mid-ocean fish include sharks and tunas. There are about 330 species of sharks. Most of these live in coastal waters. Only six species favor the mid-ocean. These include the crocodile, the whale, the blue, the shortfin mako, the oceanic whitetip, and the silvertip. Some live and travel alone; others move about in schools (large gatherings of fish). They are present in all oceans except the Antarctic. Sharks are most abundant in tropical and subtropical waters. They vary in size and behavior and are relatively free of disease. With the exception of other sharks, they have few enemies.

Some deep-ocean sharks must remain in constant movement in order to breathe. Their forward motion forces water through their gills from which they extract oxygen. Other species with more ordinary respiratory (breathing) systems can float or rest on the ocean bottom. Most sharks move leisurely, but some species are very fast. The blue shark can attain speeds of up to 40 miles (64 kilometers) per hour. A shark's sensitive eyes can see well in dim light, which enables some species to hunt in the twilight zone. They have a highly developed sense of smell and can sense movement vibrations in the water, which makes it easy for them to find prey.

All sharks are carnivorous. The rare whale shark, the largest fish in the ocean at more than 50 feet (15 meters), feeds only on small fish and plankton. The real predator is the great white, which averages 20 feet (6 meters) in length but may grow much larger. The diet of a shark includes almost every animal in the oceans, including other sharks. During feeding frenzies, large numbers of sharks attack the same prey, tearing it to pieces. In their haste to eat they may rip into one another, and any unlucky wounded shark is eaten along with the rest of the meal. Sharks often go without food for long periods, during which time oil stored in their livers helps sustain them.

Compared to some other species of fish, sharks do not produce large numbers of young. They may have as few as two or almost one hundred. Some species lay eggs, while in others, the young develop much as mammals do inside the female and are born alive. Development of the egg or embryo (baby) may take from a few months to two years,

Grey reef sharks frequent the waters near islands and coral reefs. IMAGE COPYRIGHT IAN SCOTT, 2007. USED UNDER LICENSE FROM SHUTTERSTOCK.COM.

depending on the species. The parents do not care for their young, which are completely developed at birth.

As humans probe farther into the oceans, they encounter sharks more frequently. Six species are known to attack humans: great whites, tigers, bulls, makos, hammerheads, and oceanic whitetips. In 2005 only about 58 unprovoked attacks occurred, and only a few were fatal. Most mid-ocean sharks live too far offshore to present much of a threat.

Tunas live in both temperate and tropical oceans; usually in the surface waters. They tend to be large fish, although their size varies from species to species. The largest species, the bluefin, may grow to more than 14 feet (4.2 meters) in length and weigh more than 1,400 pounds (680 kilograms). Their dark blue-green and silver coloring helps disguise them from predators. They are migrators and travel long distances in search of food and for breeding purposes. Skipjack tunas have been observed to travel as far as 16,000 miles (256,000 kilometers).

Like some species of sharks, tunas must be in constant motion in order to breathe. They are warm-bodied fish, which means their circulatory system enables them to maintain a body temperature warmer than that of the surrounding water. This gives them more power; they are fast swimmers, traveling at more than 30 miles (48 kilometers) per hour. Since their energy demands are great, they require a lot of food. They primarily eat other fish and squid.

Tunas are in great demand commercially as food for humans. For example, a bluefin tuna, which is endangered, can be worth as much as $60,000. To prevent overfishing, international agreements have been made to limit the number that can be caught and the areas where tuna fishing is allowed.

Food   Most fish live near the continental margins where food is readily available. Some are plant eaters whose diets are primarily phytoplankton, algae, or sea grasses. Fish that must swim in search of prey have more streamlined bodies than those that burrow into the bottom sediments for it.

Fish that live primarily in the twilight zone often travel to the sunlit zone at night to feed. Their large eyes aid them in finding prey, which may not be able to see as well under low-light conditions. They often use a light organ to lure prey.

Fish that live in the deepest regions must depend on food that falls from above, such as particles of zooplankton or the dead bodies of other animals. Strict carnivores are rare in the dark zone, perhaps because there is too little prey. Many dark-zone fish have bodies that stretch to accommodate large meals when they can find them. Although most dark-zone fish are not luminous, the anglerfish has a light organ used to lure prey. This organ is positioned over its mouth like the bait on the end of a fishing pole. Another fish, drawn by the light to investigate, is snapped up for a meal. Some species of dark-zone fish travel at night to higher levels for food, but they do not travel as far as fish living in the twilight zone.

Reproduction   Most fish lay eggs, although many, such as certain species of sharks, bear live young. Some, such as the Atlantic herring, abandon their eggs once they are laid. Others build nests and care for the new offspring. Still others carry the eggs with them until they hatch, usually in a special body cavity or in their mouths.

Certain marine fish, such as sturgeons and Pacific salmon, are actually born in freshwater rivers but spend most of their lives at sea, often traveling thousands of miles. Years later they return to the river where they were born. There, they breed and then die. Atlantic freshwater eels leave American and European rivers to breed in the Sargasso Sea, where they, too, die. By some inherited means of guidance, their young return to those same rivers, and the cycle is repeated.

Male anglerfish, a dark-zone species, are dwarfs. Before the breeding season, the male bites into the skin of the female, and their circulatory systems become joined so that the male receives his nutrients from the female's blood. (The circulatory system includes the heart, blood vessels, and lymphatics that carry blood and lymph throughout the body.) This may help assure that males and females can find one another at the appropriate time in this sparsely populated region. Some dark-zone fish are hermaphroditic. This means the reproductive organs of both sexes are present in one individual fish, and true mating is not necessary.

Seabirds   Most seabirds remain near land where they can nest during breeding season. A few species are known to travel great distances over the oceans in search of food and spend most of their lives doing so. The Arctic tern, for example, is the greatest long-distance migrator of all, traveling up to 20,000 miles (32,000 kilometers). These species are classified as

Albatrosses feed on fish, krill, and squid, often spending much of their lives flying over the open ocean. IMAGE COPYRIGHT ARMIN ROSE, 2007. USED UNDER LICENSE FROM SHUTTERSTOCK.COM.

oceanic birds. Many seabirds have adapted to marine environments by means of webbed feet and special glands for removing excess salt from their blood.

Common seabirds   Many seabirds remain near the coastline and spend much of their time on land. A few species, such as the albatross, petrel, and tern are oceanic birds.

Albatrosses are wanderers, spending much of their lives over the open ocean. Some appear to follow established routes; others follow the wind. Albatrosses are very large birds with long, narrow wings that may span up to 12 feet (3.5 meters) from tip to tip. In 1998, a black-footed Laysan albatross was observed to have traveled 25,135 miles (40,216 kilometers) in seventy-nine days. Albatrosses sleep on the ocean's surface, drink seawater, and feed on small fish and squid. Some species are scavengers and follow ships at sea, eating the garbage thrown overboard. Although they mate for life, the mated pair normally live thousands of miles apart and come together on land only during the breeding season.

Petrels, like albatrosses, spend most of their lives at sea. One species, called storm petrels, were once believed by sailors to signal the coming of bad weather. Storm petrels feed on plankton floating on the surface of the ocean. Some diving petrels can swim after prey underwater. Larger species eat carrion (dead animals). Like albatrosses, petrels come to land only to breed.

Terns plunge dive into the ocean to catch small fish. These sea birds encounter more hours of daylight each year than any other creature thanks to their migration patterns.

Food   All seabirds are carnivorous. Most eat fish, squid, or krill, and they live where the food they prefer is in ready supply. Several species, such as cormorants and diving petrels, hunt underwater, using their wings to swim. Other species, such as gannets, spot their prey from high above and then dive-bomb into the water, bringing their catch to the surface to eat. Some, like gulls, swoop down on fish swimming close to the surface.

Reproduction   Seabirds tend to nest on islands where there are few land-dwelling predators. Some nest in huge colonies on the ground, others dig burrows, and still others prefer ledges on a cliff. Like other birds, seabirds lay eggs and remain on the nest until the young are able to

Manatees are also known as sea cows and they spend their time grazing in shallow waters.
IMAGE COPYRIGHT WAYNE JOHNSON, 2007. USED UNDER LICENSE FROM SHUTTERSTOCK.COM.

leave on their own. Some live in one area and migrate to another for breeding. The Arctic tern, for example, spends the winter months in Antarctica, then travels halfway around the globe to breed during the summer in the Arctic Circle.

Marine mammals   The three most important mammals that are truly marine are whales, porpoises, and sea cows. They must remain in the water at all times. Other mammals, including seals and sea otters, spend much of their time in the water but are able to live on land, and do so especially at breeding time.

Common marine mammals   Whales and their cousins, porpoises and dolphins, are the primary deep-ocean mammals. Whales are found in all the world's oceans, including the Antarctic. Most species migrate, some traveling as much as 14,000 miles (22,400 kilometers) during one breeding season. Their paths usually follow ocean currents and may depend upon available food supplies and water temperature. Most species travel in schools. Smaller species may live as long as thirty years. The larger species may live to be 100 years old.

Whales fall into two categories: toothed whales and baleen whales. Instead of teeth, baleen whales have a row of bony, fringed plates they use to filter plankton from the water. Baleen whales include the gray, the humpback, and the blue whale. Blue whales are the largest animal known to have lived. They grow to lengths of 100 feet (30 meters) and may weigh as much as 200 tons (180 metric tons). Toothed whales include killer, beluga, and sperm whales. These are all predators, feeding mainly on fish and squid.

A layer of dense fat, called blubber, surrounds the whale's body directly under its skin. Whales are warm blooded and the main purpose of this fatty layer may be to help them maintain body temperature. Some species, such as the sperm whale, may dive to depths of more than 3,300 feet (1,006 meters) and stay underwater for up to an hour, but whales must return to the surface to breathe. The blowhole on the top of a whale's head is actually the nostril. The spout, which looks like a stream of water the whale shoots in the air through its blowhole, is actually exhaled air. This exhaled air is usually warmer than the outside atmosphere and its moisture quickly condenses and looks like steam.

Whales occasionally jump out of the water, landing again with a huge splash. This is called breaching. The reason for it is unknown, but it may have something to do with the fact that they are very social animals. Almost all species of whales produce sounds such as whistles, squeals, and clicks. These sounds may have much to do with social behavior, and there is little doubt that they express certain emotional states such as fear or hunger.

Food   The sea cow is the only plant-eating mammal that truly lives in the sea. Most marine mammals are carnivorous. Some, like the baleen

Whales are one of the few truly marine mammals. These are Beluga whales. IMAGE COPYRIGHT MARTIN TILLER, 2007. USED UNDER LICENSE FROM SHUTTERSTOCK.COM.

whales, feed primarily on zooplankton, especially krill. Others, like the porpoise, feed on fish and invertebrates such as squid and even clams or oysters.

Reproduction   Like other mammals, marine mammals bear live young. Most have only one offspring at a time. The young are nursed with milk produced by the mother's body until they are able to find food on their own.

DIVING ABILITIES OF DIFFERENT ANIMALS*	
Sperm whale	5,280 feet (1,609 meters)
Weddell seal	1,968 feet (600 meters)
Blue whale	1,000 feet (305 meters)
Penguin	800 feet (244 meters)
Human with aqualung	260 feet (79 meters)
Porpoise	164 feet (50 meters)

*These are distances that have been actually measured, depths may be much greater in some cases.

Endangered species   At one time sharks were plentiful; however, reduction in their food supply from commercial fishing and a growing interest in the sharks themselves for human food have reduced their numbers.

Most seabirds tend to be unafraid. Many are threatened by humans who use them for food or feathers when they come to land. On the open ocean, thousands are caught accidentally each year in fishing nets. Others suffer from pollution, pesticides, and the results of oil spills, which destroy the waterproofing effect of their feathers. The albatross population alone has declined 40 percent in the past thirty years.

Until the twentieth century and the introduction of factory whaling ships, the great whales were numerous. Beginning around 1864 and ending in the 1970s, commercial whaling reduced the numbers of many coastal species. In 1986, a number of nations suspended commercial whaling in order to allow populations to increase. Since then, a few countries allow limited hunting locally for food. The North Atlantic right whale, protected since the 1930s, has fewer than 500 specimens remaining.

The bluefin tuna has decreased to 10 percent of its population since 1975.

Human Life

People do not live in the ocean environment for long periods of time, but the oceans have had an effect on all human life. Human life has its effect on the oceans too.

Impact of the ocean on human life   Without the oceans, there would be no life on Earth. Not only did the first life forms originate there, but

Sea Serpent Sightings

For centuries, people have wondered if strange, gigantic creatures lay hidden in the unexplored regions of the sea. Every now and then, someone reports seeing what they think is a sea serpent. So far, the sightings that could be checked have turned out to be creatures already known. For example, a sea serpent that was reportedly washed up on the California coast turned out to be the remains of a beached whale. Sharks or other predators had eaten part of the whale, and its skin was twisted in such a way that it appeared to have a very long neck. Large regions of the ocean are still unexplored, and gigantic animals may yet remain to be discovered. The larva of an enormous eel has been found in the Pacific, for example. If its parent ever turns up, it might just be the legendary sea serpent.

the oceans help sustain all life forms. Oceans regulate gases in the atmosphere, and the phytoplankton that grows in them provide oxygen. The oceans influence weather and help moderate temperature; they are largely responsible for the rain, sleet, and snow that fall on land. They are also a source of food, water, energy, minerals, and metals.

Food and water Since prehistoric times, humans have depended upon the oceans for food, primarily fish. Some fish, such as orange roughy, are caught in deep waters, but most primary fishing areas are over the continental margins where the waters support more sea life.

To make up for the shortage of certain species of fish, fish farming has become popular. In fish farms, fish eggs are hatched and the young fish are fed and protected until they can be released into the ocean or sold for food. A few sea fish, such as salmon, are grown successfully in farms but their flesh lacks the high levels of nutrients found in wild salmon.

Algae are used for food and to make food products. Certain varieties of kelp are popular in Japan, where they are cultivated in sea farms and used as a vegetable. In the United States, algae are used primarily as thickeners in ice creams and puddings.

Sea salt is another food item taken from the oceans. Sea water isolated in shallow ponds along the shoreline is allowed to evaporate until a crust of crystals forms. The crystals are then collected, processed, and packaged.

Sea water is desalted and used by some desert countries for drinking and irrigation (a method of supplying water to dry land). This process, called desalination, is expensive and not very efficient. Only wealthy countries can afford to do it.

Energy Millions of years ago, sediments from dead animal and plant life (fossils) formed on the ocean floor. Time, heat, and pressure from overlying rock have worked to turn these sediments into fossil fuels, primarily gas and oil. To obtain these resources, oil rigs (large platforms standing well above sea level but anchored to the sea bed) are built. From these platforms, drilling is done into the rock, releasing the gas and oil,

Fossil fuels such as oil can be found beneath the ocean floor. Large platforms, anchored on the sea bed, are built to mount drilling equipment to retrieve these resources. IMAGE COPYRIGHT ARTHUR EUGENE PRESTON, 2007. USED UNDER LICENSE FROM SHUTTERSTOCK.COM.

which are then pumped to shore through pipelines. Most gas and oil deposits have been obtained from offshore rigs. As these deposits are used up, ways of acquiring deposits in the deep sea are being explored.

Ocean surface waters absorb large quantities of solar energy (energy from the sun). A process known as Ocean Thermal Energy Conversion has been used to capture some of that energy for human use. Conversion plants are located in Hawaii and other tropical islands. Producing usable energy from ocean currents, waves, and tides is also being explored.

Minerals and Metals　Minerals and metals are other important oceanic resources. Rocks, sand, and gravel dredged from the sea floor, especially in the North Sea and the Sea of Japan, are used in construction. Bottom deposits of manganese, iron, nickel, and copper have been found in the deep ocean, but so far, extracting them is expensive and poses environmental problems. Some minerals, such as sulfur, can be pumped as liquids from beneath the ocean floor.

Transportation　Since the first humans ventured out on the oceans in ships, the oceans have provided transportation routes. At least 40,000 years ago, people made the journey from Southeast Asia to Australia and New Guinea. Since then, seagoing routes have been used for trade, expansion, travel, and war.

Impact of human life on the oceans　The World Ocean surrounds us, so what one country does to it affects all countries. If one country overfishes a species of fish, other countries may suffer.

Wrecks at Sea

Since people first began to travel on the open sea, there have been shipwrecks. Some areas, such as the Bermuda Triangle in the Atlantic, are famous for the number of ships lost in them.

The sea helps preserve wooden ships by covering them with sediment so that wood-eating animals that might attack them are kept away. Metal ships, however, rust easily in seawater. Animals, such as corals, grow on the outside of the ship, and others, such as fish and octopi, may use the interior for shelter.

The most famous wrecked ship is the *Titanic,* which sank in 1912 on its first voyage. The *Titanic* was supposedly unsinkable because it contained many water-tight compartments that should have kept it afloat. It struck an enormous iceberg and sank in less than three hours. Of the 2,228 people on board, 1,523 died.

Use of plants and animals   After World War II (1939–45) the technology of commercial fishing improved, and a growing population increased the demand for fish as a food source. Major food species such as herring, cod, haddock, sardines, and anchovies, had been greatly reduced. Marine fishing reached an all-time high in 1989, at about 98 million tons (89 million tonnes). Regulations now limit fishing for these species. Some scientists believe that by monitoring the number of fish in a certain species, and by adding more species of those used for human food, no species should be threatened with extinction.

Fish farms are a means of helping maintain certain species of commercially popular fish. Pacific salmon are raised in hatcheries (fish farms) in the United States and Norway; oysters are raised in the United States; and shrimp farms can be found in Mexico, Ecuador, and Taiwan. Japan has greatly increased its catch of fish by building artificial reefs along its coastline. These reefs attract algae, which in turn attract fish and other sea animals.

The great whales were hunted for centuries by people of many countries for their meat, which was used for food; their fat, which was used for oil and in making soap; and for other body parts, which were ground into animal feed or used to make such items as brushes. Whales are now protected, but some scientists believe certain species will never recover from their losses and may still face extinction.

The Stellar sea cow, which looked like a giant walrus, weighed up to 11 tons (10 metric tons) and was hunted into extinction in the 1700s.

Many sea plants and animals are popular as souvenirs or art objects. When seashells from dead animals are taken, no harm is usually done. However, many shells available commercially are taken from living animals and the animals are left to die.

Marine parks and reserves have been set up all over the world to protect endangered species. They include the Shiprock Aquatic Reserve in Australia and the Hervey Bay Marine Park in California. Species on the endangered lists include the leatherback sea turtle, humpback whale, green sea turtle, and the bowhead whale.

Natural resources Large quantities of natural resources, such as oil and minerals, can be found in the oceans or beneath the ocean floor. These resources have not been used up because they are still too difficult or too expensive to obtain. As methods improve that may change.

Quality of the environment For centuries, the oceans have been used as a garbage dump. Six million tons (5,442,000 metric tons) of litter are dumped into them each year from ships alone, while sewage and industrial wastes come from coastal cities. Discarded items, such as plastic bags and old fishing nets, pose a hazard for the animals that get caught in them. Oil spills from tanker ships carrying oil from one country to another are dangerous, as is oil from oil refineries and pipelines. Efforts by concerned nations have begun to correct some of these problems. In 1972, an agreement to prohibit dumping of toxic (poisonous) materials in open seas was signed by ninety-nine nations.

At one time, radioactive nuclear wastes, which are extremely toxic, were dumped into the ocean. The Irish Sea, where this occurred, is the most radioactive sea in the world. Dumping of nuclear waste is now banned. Some countries have begun to study the deepest areas of the sea as potential disposal sites for extremely dangerous wastes, such as waste from nuclear reactors. Whether this can be done successfully without harm to the ocean environment, and ultimately to humans, remains to be seen.

Ocean exploration Humans have explored the surface of the seas since ancient times, going as far as their craft and their courage would take them. By 3200 BC the Egyptians had invented sails and were traveling by sea to different countries for trading purposes. The ocean depths are another matter; humans can go only so deep without special equipment.

During the 1600s, diving bells were designed that allowed divers to go as deep as 60 feet (18 meters). Lead-lined wooden barrels filled with air were lowered periodically to the bell and a leather tube was used to connect them with the divers. In the 1700s, compressed air (air forced into a metal container) and the development of metal helmets and flexible

---

The God of the Sea

To the ancient Greeks, Poseidon was the god of the sea. It was believed that he lived beneath the oceans's depths and controlled the fate of those who ventured out upon the waters. Poseidon could summon tsunamis, which is how he brought down the ancient kingdom of Crete, home of the fearful mythical Minotaur who was half man, half bull.

The Seven Seas

Centuries ago, people spoke of "sailing the seven seas." The seas in question were those considered navigable at the time: the Atlantic, Pacific, Indian, and Arctic Oceans; the Mediterranean and Caribbean Seas; and the Gulf of Mexico. Now scientists only speak of three major oceans, the Atlantic, Pacific, and Indian. The remainder of the original seven are considered part of the Atlantic.

diving suits made exploration easier. In 1943, Jacques Cousteau and Emile Gagnan made diving to a depth of 165 feet (50 meters) possible by perfecting the automatic aqualung. (An aqualung provides compressed air through a mouthpiece.) Diving deeper has proven difficult. The dark, the cold, and the high pressure created by the weight of the water overhead limit what humans can do without special pressurized suits and protective vehicles.

In 1960, the bathyscaphe (BATH-uh-skafe) *Trieste,* operated by the U.S. Navy, descended almost 6.8 miles (11 kilometers) into the Mariana Trench in the Pacific. A bathyscaphe is a small, submersible (underwater) vehicle that can accommodate several people and is able to withstand the extreme pressures of the deep ocean—more than 16,000 pounds (110,240 kilopascals) per square inch. Other manned submersibles, including the U.S. *Alvin,* the French *Nautile,* the Japanese *Shinkai 6500,* and the Russian *Mir I* and *Mir II,* have reached depths of 3.7 miles (6 kilometers) and have greatly added to our knowledge about the ocean floor. In 1996, the Japanese submersible *Kaiko* collected the first samples of sediment from the Challenger Deep, the lowest part of the Mariana Trench.

Safety and other considerations make manned exploration of the oceans difficult. For these reasons, remotely operated vehicles are often used for unmanned exploration. Some vehicles are about the size of a small car. They are attached by cables to a "mother" ship and may be equipped with video cameras, mechanical arms, and sensors that measure temperature, salinity, and other water conditions. New models are expected to go as deep as 13,120 feet (4,000 meters) and have cameras that can operate without lights. These newer models are not able to go as deep as previous models, like the *Trieste,* but they are more sophisticated and are meant for more elaborate studies.

Much exploration has been made using sonar equipment. Sonar is the use of sound waves to detect objects. Single pulses of sound are sent out by a machine at regular intervals and as the sound pulses are reflected back a "picture" is obtained of the surrounding area. Sonar has helped researchers map the mountains and valleys of the ocean floor.

The Food Web

The transfer of energy from organism to organism forms a series called a food chain. All the possible feeding relationships that exist in a biome make up its food web. In the ocean, as elsewhere, the food web consists of producers, consumers, and decomposers. These three types of organisms transfer energy within the oceanic environment.

Phytoplankton are the primary producers in the oceans. They produce organic materials from inorganic chemicals and outside sources of energy, primarily the sun.

Zooplankton and other animals are consumers. Zooplankton that eat only plants are primary consumers in the oceanic food web. Secondary consumers eat the plant-eaters. They include the baleen whale and zooplankton that eat other zooplankton. Tertiary consumers are the predators, like tunas and sharks. Humans are also tertiary consumers called omnivores, organisms that eat both plants and animals.

Decomposers feed on dead organic matter and include lobsters and large petrels. Bacteria help in decomposition.

Harmful to the oceanic food web is the concentration of pollutants and dangerous organisms. It was once thought that the ocean would dilute harmful chemicals, but just the opposite is true. They become trapped in sediments where life forms feed. These life forms are fed upon by larger organisms, and at each step in the food chain the pollutant becomes more concentrated. When humans eat contaminated sea animals, they are in danger of serious illness. The same is true of diseases such as cholera, hepatitis, and typhoid, which can survive and accumulate in certain sea animals and then be passed on to people.

Spotlight on Oceans

The Indian Ocean   The Indian Ocean is the third largest in the world and covers about 20 percent of Earth's water surface. Its volume is estimated to be about 62,780,380 cubic miles (261,590,400 cubic kilometers).

The Indian Ocean contains many islands. During the prehistoric breakup of the continents, small pieces of continent were left behind in the Indian Ocean as undersea plateaus (high, level land areas). Some of these plateaus rise above the water and form islands, such as the Laccadives. Many other islands, such as Mauritius, are volcanic in origin. On the ocean's eastern border lie the islands of Indonesia; on the western

The Indian Ocean
Location: South of India, Pakistan, and Iran; east of Africa; west of Australia; north of the Antarctic Sea
Area: 28,000,000 square miles (73,000,000 square kilometers)
Average Depth: 12,760 feet (3,890 meters)

border lie Madagascar, Zanzibar, the Comoros, the Seychelles, the Maldives, and the Nicobar Islands. To the south are the Crozets and the Kerguelen. Coral reefs (undersea walls made from coral skeletons) can be found in areas of the ocean located in the tropics.

Running through the center of the Indian Ocean's floor is part of the mid-ocean ridge in the form of an upside-down Y. Many peaks in this mountain chain are about 6,560 feet (2,000 meters) high. All along the chain are active volcanoes; earthquakes constantly occur, causing spreading of the sea floor. The Indian Ocean basin is expanding at a rate of about 1 inch (2.5 centimeters) each year.

The Java Trench is the only known trench in the Indian Ocean. It lies south of Indonesia and, at its lowest point, is 4.5 miles (7 kilometers) deep. To the north of the trench is another string of volcanoes, the most famous of which is Krakatoa, which exploded so violently in 1883 that it could be heard 1,860 miles (3,000 kilometers) away.

Several major rivers flow into the Indian Ocean bringing large quantities of sediment. These include the Indus and Ganges in India and the Zambesi in southern Africa. Over thousands of years, these sediment layers have formed vast fans that spread out over the nearby ocean floor.

Climatic conditions over most of the Indian Ocean are tropical. In its warmest part, the Arabian Sea, surface waters can reach 86°F (30°C). Near the Antarctic Sea, the temperature can drop to less than 54°F (12°C). Average annual rainfall is about 40 inches (102 centimeters). In the ocean's northern reaches, rainfall is affected by the monsoons (rainy seasons) that drench the Asian continent. In the southern portion, trade winds blow from the southeast all year long.

Two major currents occur in the Indian Ocean. In the southwest, the Agulhas flows between Africa and Madagascar. At the equator, the North Equatorial Current occurs in the winter and flows west.

Where the Indian Ocean meets the Pacific Ocean in the region of the Philippines, marine life is very rich. In the open ocean marine life is scarce because the waters of the Indian Ocean are warm, and growth of phytoplankton is limited. As a result, the creatures that feed on phytoplankton are limited.

Among the species of invertebrate plankton that live in the Indian Ocean is the sea wasp jellyfish. Its poisonous sting produces large welts. A person with many stings can die in minutes. Crabs, lobsters, oysters, squid, and giant clams are found here.

The Indian Ocean supports many species of seabirds, especially around the shoreline where food is plentiful. The most common birds are noddies, boobies, terns, frigate birds, storm petrels, and albatrosses.

Species of mid-ocean fish include sharks, flying fish, tunas, marlins, and sunfish. The coelacanth, a species of fish surviving from prehistoric times, has been found off the Comoros Islands, where it is now protected.

Whales are plentiful in the Indian Ocean, especially in the cooler southern waters where food is abundant. Much of the Indian Ocean is now a protected area for whales. Other mammals include fur seals, elephant seals, and, in northern portions, sea cows. Sea cows have become endangered because they are easily caught in nets and killed for their meat and hides.

Commercial fishing in the Indian Ocean is limited to local needs. Over the centuries, the ocean's greatest value has been for trade transport. Since the discovery of large oil deposits in the Middle East, it has been key to the shipment of petroleum extracted along the Persian Gulf.

The first Westerner to explore the Indian Ocean was Vasco da Gama (c. 1460–1524) of Portugal. In May 1498, da Gama reached India and, for the next century, Portugal claimed this ocean as part of its empire. The ocean was so vast that no one nation controlled the surrounding lands until England in the early 1800s. After World War II (1939–45), England withdrew from the area. Gradually, India, Russia, and the United States have become major influences. Countries bordering the ocean want it declared a peaceful zone where all people may travel the waters freely and safely.

The Sargasso Sea

The Sargasso (sar-GAS-oh) Sea is a clear, saucer-shaped area of water near the island of Bermuda in the Atlantic. It is formed by two main opposing ocean currents, the Gulf Stream to the north, and the North Equatorial Current to the south. Its waters are warm and, because of the action of the currents, they slowly revolve clockwise above much colder Atlantic depths. This rotation causes the water in the center to rise, and the level of the Sargasso is about 3 feet (1 meter) higher than the water surrounding it.

The name Sargasso comes from Sargassum, the type of brown algae that grows there in abundance. The algae are plankton, and they possess clusters of gas-filled chambers resembling grapes at the bases of their fronds that keep them afloat. Huge masses of the algae drift on the surface of the sea. These algae reproduce when pieces break off the main

The Saragasso Sea
Location: Western North Atlantic
Area: 2,000,000 square miles (5,200,000 square kilometers)
Average Depth: 3 miles (4.8 kilometers)

organism and begin to grow. Every piece can potentially grow into a new organism.

Many animals live in the Sargasso Sea that are ordinarily not found in mid-ocean because of the plentiful algae. Some animals, like tubeworms, attach themselves to the algae and sift the water for tiny organisms they use for food. Crabs, shrimp, and snails roam everywhere over the fronds. The leptocephalus (lep-TOH-sef-a-LOS) eel migrates to the Sargasso to breed but otherwise lives in freshwater rivers thousands of miles away.

Permanent species of fish include the sargassum fish, which lives only in the Sargasso. Its scientific name means "the actor;" an appropriate name because it spends its life pretending to be a sea weed frond. Its body has black and yellow-green blotches to match the algae. It has a pair of fins with fingerlike projections that allow it to attach to a frond and drift as the frond does. When it stalks its prey, it climbs over the weeds. One of its fingerlike projections resembles a bit of food, which it dangles in front of its mouth waiting for another fish to take the "bait."

Italian explorer Christopher Columbus (1451–1506) reported on the Sargasso after his first voyage to the "Indies," and he claimed to have found evidence of other voyagers there. It is possible that the Carthaginians, who lived in the city of Carthage on the North coast of Africa during ancient times, may have reached the Sargasso as early as 530 BC. Legends tell of ships being trapped in the weeds, but that likely never happened. Instead, ships set adrift may have been carried here because of the rotating current, and this may be the source of the myth.

The Black Sea   The Black Sea is the world's largest inland body of water. It qualifies as a sea because it remains connected to the Sea of Marmara, the Aegean Sea, and finally to the Mediterranean Sea by means of the Bosporus Strait. Large European rivers including the Dnieper, the Danube, the Dniester, and the Don flow into it. The fresh water they bring makes it less salty than most seas.

Two currents flow through the 19-mile (30-kilometer) -long Bosporus Strait. A rapid surface current carries water from the Black Sea toward the Aegean and eventually into the Mediterranean. Beneath this current, a strong undercurrent travels in the opposite direction, bringing waters from the Mediterranean into the Black Sea. These currents create very choppy waters and help create the Black Sea's unusual environment.

The water column in the Black Sea is layered. Salt water coming from the Mediterranean enters at a deep level and continues to sink. Fresh

The Black Sea
Location: North of Turkey, south of Russia and the Ukraine
Area: 162,000 square miles (419,580 square kilometers)
Average Depth: 3,826 feet (1,166 meters)

water from the rivers flows into the shallow coastal areas. Fresh water is lighter than salt water, so it floats on top of the salt water. The two layers mix very little. The bottom layer, which consists of almost 90 percent of the water column, receives little oxygen creating a dead zone. Nothing lives at the bottom of the Black Sea except a few species of bacteria. A form of sulfur dissolved in the deep water gives it the odor of rotten eggs.

The upper layer supports abundant life forms. About 300 species of algae live in the upper layer to a depth of about 65 feet (20 meters). These plants provide food for zooplankton, mollusks, and other sea life. Many kinds of fish live in the upper layer, including anchovies, bluefish, turbot, and sturgeon. The largest mammals in the Black Sea are dolphins. More than 1 million dolphins once lived there, but their numbers were greatly reduced by commercial fishing. Some countries have begun programs to protect the dolphins and other endangered species.

Countries surrounding the Black Sea once depended upon it financially for fish. Overfishing, the use of pesticides, industrial pollution, diversion of river waters for irrigation, and nuclear contamination from the Chernobyl reactor explosion in 1986 have caused fish populations to decrease. Lack of fresh river water has resulted in an increase in the size of the dead zone, which may eventually take over the entire Black Sea. If that happens, nothing will be able to live in it except bacteria.

In Greek legend, the Black Sea was the body of water on which Jason and the Argonauts sailed in search of the Golden Fleece. For centuries, the Black Sea was considered unfriendly for sailors because of its sudden storms and strong currents. These dark storms and heavy threatening fogs may have been what earned the sea its name.

The Pacific Ocean The Pacific Ocean is the world's largest ocean. Waters in its northern and southern halves seldom mix. In the north it is linked to the Arctic Sea by means of the Bering Strait. In the south it is bordered by the Antarctic Sea. Its volume is about 154,960,672 cubic miles (643,375,552 cubic kilometers).

The mid-ocean ridge cuts through the Pacific basin from Japan to Antarctica and attains a height of 13,000 feet (3,962 meters) in some places. Trenches along the continents often exceed 26,000 feet (7,925 meters) in depth. Deeper trenches are found along strings of islands, such as the Aleutians and the Philippines. The deepest trench in the world, the Mariana, is found in the Pacific near the Mariana Islands.

The Pacific Ocean
Location: West of North and South America; east of Asia and Australia
Area: 64,000,000 square miles (1,666,000,000 square kilometers)
Average Depth: 12,700 feet (3,870 meters)

Islands are numerous in the Pacific, and most have been created by volcanoes. The famed "Ring of Fire," an area of intense volcanic activity is found in the region of Indonesia. Islands near the equator usually have coral reefs.

Around the equator, the trade winds maintain a permanent current moving from east to west. As this current turns northward around the island of Japan, it is called the Kuroshio (koo-ROH-shee-oh) current. Like the Gulf Stream in the Atlantic, the Kuroshio is a strong, intense current.

The areas of the Pacific most abundant with plant and animal life are in the far north and far south where icy water circulating upward from the sea floor brings nutrients to the surface. Millions of tons of phytoplankton and zooplankton drift upon the waters in spring and summer, providing food for baleen (toothless) whales and basking sharks, as well as smaller animals. Species from one area seldom inhabit the other.

The Pacific is home to one of the most dangerous of the invertebrates, the little blue-ringed octopus, whose sting is highly poisonous and usually deadly. Another invertebrate, the Pacific lobster, lacks claws and uses its antennae instead for defense.

Several species of sharks and rays are found only in the Pacific. The Port Jackson shark eats primarily clams and other mollusks, using its powerful jaws to crack the shells. This particular shark has a long history; its form has remained unchanged for the past 150 million years. More than twenty species of whales, porpoises, and dolphins live only in the Pacific. In the southern regions, leopard and fur seals can be found.

More than 40 percent of commercial catches of fish with fins, such as anchovies and tuna, come from the Pacific. Other Pacific resources with commercial importance include offshore deposits of iron ore near Japan, and tin near Southeast Asia.

The Atlantic Ocean   The Atlantic Ocean contains about 25 percent of all the water in the World Ocean. Its volume is 73,902,416 cubic miles (307,902,776 cubic kilometers).

Although its waters are typically less salty than those of the Pacific, its northern portion is the warmest and saltiest area in the World Ocean. This is due, in part, to water flowing into it from the Mediterranean. Its circulation is limited because the northern portion is hemmed in by continents, and because its waters do not readily mix with those of the Arctic Ocean to the north. Currents are stronger than in the southern Atlantic.

The Atlantic Ocean
Location: East of North and South America, west of Europe and Africa
Area: 33,000,000 square miles (86,000,000 square kilometers)
Average Depth: 12,100 feet (3,688 meters)

The mid-ocean ridge forms an S-shape in the Atlantic basin and divides it into two parallel sections. Some peaks along the ridge form islands, such as the Azores. Its deepest trench is the Puerto Rico with a depth of 28,232 feet (8,605 meters).

The most familiar Atlantic current is the Gulf Stream, which flows along the eastern coast of North America. A number of currents in the North Atlantic form a clockwise gyre. In the South Atlantic, currents form a counterclockwise gyre.

The Atlantic is geologically young, and its bottom-dwelling animals are descendants of animals that migrated there from other oceans. As a result, the numbers of bottom-dwelling species are few. In its equatorial and temperate regions, nektonic and planktonic animals are abundant. Many ocean travelers, such as sharks, whales, and sea turtles, cross the southern regions when they migrate.

Atlantic resources include diamonds found off the southwest coast of Africa, sand and gravel off the coast of northwest Europe, and oil and gas in the Caribbean and North Sea regions. Fish taken commercially from the Atlantic include salmon, lobster, shrimp, crabs, sardines, and anchovies.

The science of oceanography began in the northern Atlantic, and many theories about oceans are based on studies made there. The northern portion of the Atlantic differs from its southern portion and from the Indian and Pacific Oceans, so these generalizations cannot be automatically accepted.

The Antarctic Ocean The Antarctic Ocean is sometimes called the Southern Ocean. It encircles the continent of Antarctica, and in winter ice forms over more than 50 percent of the ocean's surface. Glaciers break free of the continent periodically and drift seaward to join floating ice shelves.

In the north, the mid-ocean ridge borders the Antarctic basin. The basin itself reaches a depth of about 18,400 feet (5,600 meters).

Currents in the Antarctic Ocean flow from west to east and the water is turbulent (rough). The area is windy and usually cloud covered.

The amount of daylight varies dramatically at the south pole. In winter, it is dark almost all of the time, and light almost all of the time in summer. Growth of phytoplankton is limited to spring and summer seasons.

The most important zooplankton is krill, which is food for baleen whales, seals, sea birds, fish, and squid. Among invertebrates, squid are

The Antarctic Ocean
Location: Surrounding the continent of Antarctica
Area: 12,451,000 square miles (32,249,000 square kilometers)
Average Depth: 12,238 feet (3,730 meters)

common and play an important part in the food web. They are eaten by sperm whales and albatrosses.

Most fish in the Antarctic are bottom-dwellers, where they tend to live on or near the continental margins. Many, such as the Antarctic cod, contain a kind of antifreeze in their body fluids that allows them to live at temperatures below freezing.

Sea birds, such as penguins, breed on the continent. The Adelie penguin is the most numerous. Breeding colonies are usually densely packed because ice-free land is limited.

Among mammals, baleen whales, toothed whales, and seals thrive there. Overhunting in the area has threatened many species, especially the blue whale, and it is not known if some of these species will survive. In 1982, the Convention for the Conservation of Antarctic Marine Living Resources was agreed to by concerned nations in order to better regulate fishing in the Antarctic Ocean.

The Red Sea   The Red Sea lies in the heart of the Middle East and is surrounded by desert. In the south it is connected to the Gulf of Aden and the Indian Ocean by means of the Straits of Bab el Mandeb. In the north, it is linked to the Mediterranean Sea by means of the Gulf of Aqaba and the Gulf of Suez.

The Red Sea started to form 40 million years ago when the area that is now Saudi Arabia broke away from the continent of Africa. The sea floor here is still spreading at a rate of 0.5 inch (1 centimeter) per year, and scientists consider the Red Sea a "baby" ocean that will, after about 200 million years, be as large as the current Atlantic Ocean.

The presence of surrounding desert lands causes a high degree of evaporation in the Red Sea making its water very salty. No fresh water is added because no rivers drain into it. Another source of its saltiness is the presence of hot salty pools approximately 1.3 miles (2.1 kilometers) below its surface. It is believed that the salt in these pools comes from sediments found in some parts of the Red Sea. These sediments are rich in iron, manganese, zinc, and copper, and may prove very valuable when methods for mining them become practical.

In summer, winds come from the northwest and currents flow toward the Gulf of Aden. In winter, a southeasterly wind causes the surface currents to flow in the opposite direction.

The Red Sea
Location: East of Egypt, Sudan, and Ethiopia; west of Saudi Arabia and Yemen
Area: 169,000 square miles (437,700 square kilometers)
Average Depth: 1,730 feet (524 meters)

Normally, the Red Sea is a bluish-green color. However, it is heavily populated by orange colored algae that turn reddish-brown when they die. This may be why it was named the Red Sea.

Over 400 species of corals are found here. As the corals form reefs, they attract the types of fish that thrive in those areas. The largest fish that live in the Red Sea are whale sharks. Manta rays are also common. Other fish include snappers, grouper, parrotfish, and sardines.

Animals found in the Red Sea are important to human life. Sea cucumbers, a sausage-shaped invertebrate, are common. They are a Far Eastern delicacy making them commercially important, as are prawns, a type of tropical shrimp. Pearls from oysters in the Red Sea are famous for their high quality. Mother-of-pearl, taken from the shell of another sea animal, the mollusk, is used to make shirt buttons and other decorations.

The process of desalting ocean water to produce fresh water is becoming more common in countries bordering the Red Sea. There are eighteen desalination factories along the Saudia Arabian border. These factories remove salt from water to make it drinkable.

The Red Sea has been important for transportation since at least 2000 BC. Until 1869, the only access to the Red Sea was from the south; it was closed at its northern end until the Suez Canal was constructed by the British. The Suez Canal is a long channel that allows the waters of the Red Sea to mix with those of the Mediterranean. Ships going through the canal can reach the Indian Ocean without traveling all the way around the continent of Africa, a great savings of time and distance.

The Red Sea is made famous from the Bible because God gave Moses the power to part its waters. Moses was then able to lead the Israelites across the sea floor during their escape from bondage in Egypt.

The Mediterranean Sea The Mediterranean Sea is the world's largest inland sea. It lies between the continents of Europe and Africa. It is connected to the Atlantic in the west by the Straits of Gibraltar and to the Red Sea in the east by the Suez Canal. To the north, it joins the Aegean and Black Seas, which are also inland seas.

The Mediterranean is bisected by a submarine ridge into eastern and western basins. Many of its islands were formed by volcanoes, some of which are still active. The region has many earthquakes. Its greatest depth, 16,814 feet (5,125 meters), is in the Matapan Trench.

The climate around the Mediterranean is warm and wet in the winter and hot and dry in the summer. Surface water temperatures range from

The Mediterranean Sea
Location: Between Europe and North Africa
Area: 1,145,000 square miles (2,965,500 square kilometers)
Average Depth: 4,902 feet (1,494 meters)

about 41°F (5°C) in the north during the winter to 88°F (31°C) in the south during the summer. Many rivers flow into the Mediterranean, the largest of which is the Nile River of Egypt, but its water is very salty because of continual evaporation.

The warm water means less phytoplankton, and less phytoplankton means fewer animal species in general. The sea varies in temperature, salinity, and depth from place to place, and a variety of plant and animal life can be found. Invertebrates include crabs, lobsters, shrimp, oysters, mussels, clams, and squids. More than 400 species of fish are found there including bass, flounder, tuna, sharks, and mackerel.

Since ancient times, the Mediterranean has been a source of fish for people living in the surrounding countries. However, the catches are not large enough to be commercially important worldwide. Of greater economic importance is the discovery of oil and gas deposits beneath the sea floor.

The Mediterranean suffers from pollution as a result of industrial and municipal wastes dumped along the European coast. Oil tankers travel its waters and oil spills add to the problem. Efforts are being made to repair the damage.

The Mediterranean has been historically important since the Egyptian people began to explore it as early as 3000 BC. In later millennia, Crete, Greece, and Anatolia began to use it for trade and for ships of war. Between 300 BC and 100 AD, Rome ruled the sea. After the fall of Rome in 476 AD, the Arabs, Germans, Slavs, and Ottoman Turks each took a turn holding sway over the area. During the 1700s, the discovery of new sea routes to India around Africa made the Mediterranean less important for travel and commerce.

For More Information

BOOKS

Ballesta, Laurent. *Planet Ocean: Voyage to the Heart of the Marine Realm.* National Geographic, 2007.

Carson, Rachel L. *The Sea Around Us.* Rev. ed. New York: Chelsea House, 2006.

National Oceanic and Atmospheric Administration *Hidden Depths: Atlas of the Oceans.* New York: HarperCollins, 2007.

Ocean. New York: DK Publishing, 2006.

Sverdrup Keith A., and Virginia Armbrust. *An Introduction to the World's Oceans.* Boston: McGraw-Hill Science, 2006.

Worldwatch Institute, ed. *Vital Signs 2003: The Trends That Are Shaping Our Future.* New York: W.W. Norton, 2003.

PERIODICALS

"Arctic Sea Ice Reaches Record Low." *Weatherwise.* 60. 6 Nov-Dec 2007: 11.

Bischof, Barbie. "Who's Watching Whom?" *Natural History.* 116. 10 December 2007: 72.

Haedrich, Richard L. "Deep Trouble: Fishermen Have Been Casting Their Nets into the Deep Sea After Exhausting Shallow-water Stocks. But Adaptations to Deepwater Living Make the Fishes There Particularly Vulnerable to Overfishing—and Many are Now Endangered." *Natural History.* 116. 8 October 2007: 28.

Tucker, Patrick. "Growth in Ocean-current Power Foreseen: Florida Team Seeks to Harness Gulf Stream." *The Futurist.* 41. 2 March-April 2007: 8.

ORGANIZATIONS

Center for Marine Conservation, 1725 De Sales St. NW, Suite 600, Washington, DC 20036, Phone: 202-429-5609 Internet: http://www.cmc-ocean.org.

Environmental Defense Fund, 257 Park Ave. South, New York, NY 10010, Phone: 212-505-7100; Fax: 212-505-2375, Internet: http://www.edf.org.

Envirolink, P.O. Box 8102, Pittsburgh, PA 15217, Internet: http://www.envirolink.org.

Environmental Protection Agency, 401 M Street, SW, Washington, DC 20460, Phone: 202-260-2090, Internet: http://www.epa.gov.

Friends of the Earth, 1717 Massachusetts Ave. NW, 300, Washington, DC 20036-2002, Phone: 877-843-8687; Fax: 202-783-0444; Internet: http://www.foe.org.

Greenpeace USA, 702 H Street NW, Washington, DC 20001, Phone: 202-462-1177; Fax: Internet: http://www.greenpeace.org.

Sierra Club, 85 2nd Street, 2nd fl., San Francisco, CA 94105, Phone: 415-977-5500; Fax: 415-977-5799, Internet: http://www.sierraclub.org.

World Meteorological Organization, 7bis,avenue de la Paix, Case Postale No. 2300 CH-1211PO Box 2300, Geneva 2, Switzerland, Phone: 41 22 7308111; Fax: 41 22 7308181, Internet: http://www.wmo.ch.

World Wildlife Fund, 1250 24th Street NW, Washington, DC 20090-7180, Phone: 202-293-4800; Internet: http://www.wwf.org.

WEB SITES

CBC News Indepth: Oceans: http://www.cbc.ca/news/background/oceans/part2.html (accessed on September 1, 2007).

Distribution of Land and Water on the Planet: http://www.oceansatlas.com/unatlas/about/physicalandchemicalproperties/background/seemore1.html (accessed on September 1, 2007).

The Natural History Museum: Fathom: Deep Ocean: http://www.fathom.com/course/10701050/sessions.html (accessed on September 1, 2007).

National Oceanic and Atmospheric Administration: http://www.noaa.gov (accessed on September 1, 2007).

The Ocean Environment: http://www.oceansatlas.com/unatlas/about/physicalandchemicalproperties/background/oceanenvironment.html (accessed on September 1, 2007).

Rain Forest

A tree is a woody plant with a single, strong trunk and many branches that lives year after year. A large group of trees covering at least 25 percent of the area where the tops of the trees, called crowns, interlock to form an enclosure, or canopy, at maturity make up what is called a forest. The term rain forest is used to refer to any forest in tropical or semitropical regions. (Tropical regions are those around the equator.) Rain forests occur in a few regions with temperate (moderate) climates.

Tropical rain forests, also called jungles, are located for the most part in the belt between the tropics of Cancer and Capricorn. The term jungle usually indicates a disturbed, tangled, tropical rain forest with vines and other distinct plant and animal life. Plenty of rain and warm temperatures year-round support constant plant growth and great diversity of species. It is estimated that there are up to 260 different kinds of trees per square mile of rain forest. Trees that would normally lose their leaves during the autumn season in cooler climates retain them for several years (some trees for up to sixteen years), becoming evergreens because there are no cold and warm seasons. Rain forests cover only about 7 percent of Earth's surface, or 4.4 million square miles (11.5 million square kilometers) of land.

Tropical evergreen rain forests occur in four main regions:

- The Americas and the Caribbean
- Africa and Eastern Madagascar
- India and Malaysia
- Australia

How Rain Forests Develop

The first forests evolved during Earth's prehistoric past. Since then, all forests have developed in essentially the same way, by means of a process called succession.

WORDS TO KNOW

Angiosperms: Trees that bear flowers and produce seeds inside a fruit; deciduous and rain forest trees are usually angiosperms.

Clear-cutting: The cutting down of every tree in a selected area.

Elfin forest: The upper cloud forest at about 10,000 feet (3,000 meters) which has trees that tend to be smaller, and twisted.

Emergents: The trees that stand taller than surrounding trees.

Epiphytes: Plants that grow on other plants with their roots exposed to the air. Sometimes called "air" plants.

Forbs: A category of flowering, broadleaved plants other than grasses that lack woody stems.

Gymnosperms: Trees that produce seeds that are often collected together into cones; most conifers are gymnosperms.

Rhizomes: Plant stems that spread out underground and grow into a new plant that breaks above the surface of the soil or water.

Tannins: Chemical substances found in the bark, roots, seeds, and leaves of many plants and used to soften leather.

Understory: A layer of shorter, shade-tolerant trees that grow under the forest canopy.

The first forests   The first forests were very different from those found today. They were mostly composed of huge ferns and clubmosses; there were no flowers to speak of. Flowering plants developed only 65 to 145 million years ago during the Cretaceous period. Among the first to evolve were the ancestors of modern water lilies. Present day rain forests began developing shortly after the flowering plants.

Succession   Trees compete with one another for sunlight, water, and nutrients, thus a forest is constantly changing. The process by which one type of plant or tree is gradually replaced by others is called succession. Succession can occur naturally when different species of trees become dominant as time progresses and the environment changes. It can also occur from natural disasters, such as forest fires.

Primary succession   Primary succession of some forests in North America usually begins on bare soil or sand where no plants grew before. When the right amount of sunlight, moisture, and air temperatures is present, seeds begin to germinate (grow). These first plants are usually made up of the grasses and forbs (a nonwoody broad-leaved plant) type. They continue to grow and eventually form meadows. Over time, and as conditions change,

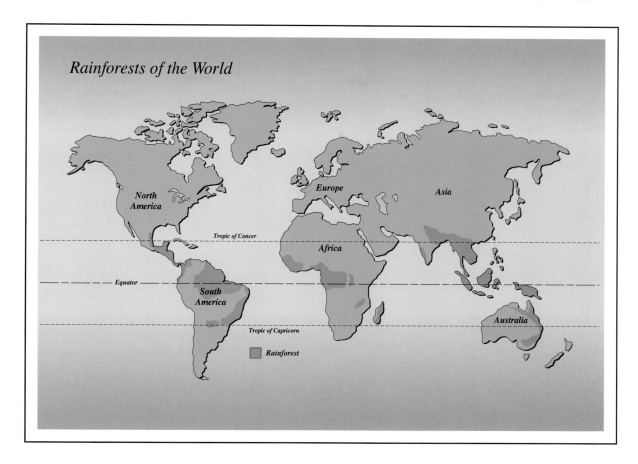

Rainforests of the World

other plants begin to grow such as shrubs and trees, which become dominant and replace or take over where the grasses and forbs originally grew.

As primary succession continues, "pioneer" trees that are tall and sun-loving quickly take over the meadow. They change the environment by making shade. Trees with broader leaves that prefer some protection from the sun can then take root. If conditions are right, a mixed forest of sun-loving and shade-loving trees may continue for many years. Eventually, more changes occur.

The climax forest   Seedlings from pioneer trees do not grow well in shade; therefore, new pioneer trees do not grow. As the mature trees begin to die from old age, disease, and other causes, the broad-leaved trees become dominant. The shade from these broad-leaved trees can be too dense even for their own seedlings. As a result, seedlings from trees

One view in the Monteverde Cloud Forest in Costa Rica. IMAGE COPYRIGHT STEFFEN FOERSTER PHOTOGRAPHY, 2007. USED UNDER LICENSE FROM SHUTTERSTOCK.COM.

that prefer heavy shade begin to thrive and eventually dominate the forest. These trees produce such deep shade that only trees or plants that can survive in complete shade will succeed there. When this happens, the result is a climax forest.

Few true climax forests actually exist because other changes take place that interfere with a forest's stability. Fires, floods, high winds, and people can all destroy a single tree to acres of trees. Glaciers can mow them down; volcanoes can smother them with ash or molten rock or knock them over with explosive force. Then the process of succession starts over.

Secondary succession When the land has been stripped of trees, in some areas it will eventually be covered with them again if left alone. This is called secondary succession and can take place more quickly than primary succession. Seeds from other forests in neighboring regions are blown by the wind or carried by animals to the site. The seeds take root, seedlings sprout, and the process begins again.

Kinds of Rain Forests

Tropical evergreen rain forests can be classified into three main types: lowland, montane, and cloud.

Lowland wet forest Lowland wet forests thrive close to the equator where rainfall is heavy, at elevations from sea level up to 4,000 feet (1,200 meters). They occur widely in the Asian tropics, in South America near the Amazon, and in Africa. Vegetation is wide and varied in these forests, having as many as 2,000 different species of trees. The largest lowland wet forest is currently located in Brazil (commonly known as *Selva*, the Spanish word for forest). Studies show that the Amazon forest may contain up to 120 different species in one acre of forest.

Montane rain forest The true montane, or mountain, rain forests begin where lowland wet forests leave off; about 4,000 feet above sea level.

There are lower and upper montane forests at lower and higher altitudes respectively. The average temperature is 66°F (19°C), and oak and laurel trees are predominant. At higher altitudes (upper montane) the cooler weather is favored by coniferous trees and myrtles. Montane forests can be found in Africa, Papau New Guinea, and South America.

Cloud forest   Cloud forests, such as those in Costa Rica, grow on mountains but usually at altitudes higher than 5,000 feet (1,500 meters). Their name comes from the fact that low-lying clouds form around them, shrouding them in mist. At about 10,000 feet (3,000 meters), the upper cloud forest is known as the elfin forest. The trees, including some species of pines, tend to be stunted and more twisted than those at lower elevations. Mosses, ferns, and lichens are abundant throughout this forest.

Climate

The climate of a tropical rain forest is warm and humid all year long. Average annual temperatures are between 68° and 84°F (20° and 29°C), and the temperature never drops below 64°F (18°C). There is almost no seasonal temperature change because the difference between the coldest and warmest months is only 3°F (1°C).

Rainfall varies throughout the tropics, with some areas receiving as little as 60 inches (1.5 meters) annually, and others twice that amount. Dry seasons occur, especially in monsoon climates. Some precipitation (rain) falls almost every day. It may come in a thunderous downpour or in a misty shower. In general, humidity is very high, and days are often cloudy.

Geography of Rain Forests

The geography of rain forests varies, depending upon location, and includes landforms, elevation, soil, and water resources.

Landforms   The terrain over which lowland forests grow features valleys, rolling hills, old river basins, and level areas. Montane forests develop in mountainous regions, as do cloud forests. Tropical mountains tend to be volcanic in origin, and their slopes are often gentle rather than steep and craggy (rugged and uneven).

Elevation   Rain forests grow at almost all elevations, from sea level to about 10,000 feet (3,000 meters). The greatest share of rain forested area exists at

Cecropia trees grow in the lower canopy layer of the rainforest and can grow six feet in height each year. IMAGE COPYRIGHT DR. MORLEY READ, 2007. USED UNDER LICENSE FROM SHUTTERSTOCK.COM.

elevations below 4,000 feet (1,200 meters), in what are known as lowland wet forests.

Soil   Rain forest soil tends to be red or yellow in color and low in nutrients because the vast number of plants quickly absorb its valuable minerals. Decaying vegetation continually enriches the topsoil, which is the most fertile soil near the surface. For this reason, tree roots are more likely to remain near the surface. In forests where the shade is dense, few smaller plants may grow and the topsoil layer is even richer. Where the ground is rocky and not much soil is present, trees may develop stiltlike projections from their trunks that help anchor them to the ground.

The presence of volcanoes increases the richness of soil in Indonesia and parts of Central and South America. Volcanic ash contains many minerals and adds nutrients to the soil.

Water resources   In tropical regions, rivers and streams are often the primary water resources. Daily rainfall not only helps maintain them, but can also cause them to flood.

Plant Life

Most forests contain a mixture of many different species of trees, and rain forests contain the most species. Both coniferous (cone-bearing) and nonconiferous evergreen trees exist within their boundaries. Exploration of most rain forests is challenging for humans so many of these species are yet to be named.

The trees and smaller plants in a rain forest grow to different heights, forming layers. The crowns of tall trees create a canopy, or roof, over the rest of the vegetation that averages 120 feet (37 meters) above the ground. The very tallest trees, called emergents, pop through the canopy like lonely towers, some as tall as 200 feet (61 meters). Trees help support one another because they grow so close together. Beneath the canopy grows at least one more layer of shorter, shade-tolerant trees, which cover heights of 30 to 65 feet (10 to 20 meters). This shorter layer is called the understory. The next layer, about 16 feet (3 meters) off the ground, is composed of tree seedlings, a few small shrubs, and a few flowering plants. The very lowest layer consists of small plants, like mosses, that live atop the soil.

Non-tree plants that grow in the rain forest are often either climbers, epiphytes (EPP-ih-fites), or parasites. Climbers have roots in the ground, but use hooklike tendrils to climb up the trunks and along the limbs of trees in order to reach the canopy where there is light. Epiphytes, or "air" plants, store water in their fleshy stems and leaves. They also grow on trees and other plants, especially in the canopy, but their roots are exposed to the air. These plants absorb the nutrients they need from rain and forest debris. Parasites attach themselves to other plants and trees, but they manage to do without light and take their nourishment from their host.

Plant life within the rain forests includes not only trees but also algae (AL-jee), fungi (FUHN-ji), lichens (LY-kens) and green plants.

Algae, fungi, and lichens   Algae, fungi, and lichens do not fit neatly into either the plant or animal category.

Algae   Most algae are single-celled organisms, although quite a few are multicellular. Most algae have the ability to make their own food in a process called photosynthesis (foh-toh-SIHN-thuh-sis). During this process they use the energy from sunlight to change water and carbon dioxide into the sugars and starches they use for food. Other algae absorb nutrients from their surroundings.

Although most algae are water plants, green and blue-green algae do appear in the rain forest where they encrust the leaves of trees. This blocks the sunlight from the trees' leaves, but the green-blue algae may aid the tree in obtaining nutrients, such as nitrogen, from the atmosphere.

Fungi   Unlike algae, fungi cannot make their own food by photosynthesis. Some fungi, like molds and mushrooms, obtain nutrients from dead or decaying organic matter (material derived from living organisms). They assist in the decomposition (breaking down) of this matter and in releasing the nutrients needed by plants back into the soil.

Rotten Flower

The Rafflesia (rah-FLEA-e-ah) plant of Malaysia is a parasite. Its seeds burrow beneath the bark of another plant and invade it with hairlike strands that absorb nutrients. Eventually a flower bud emerges, but the bud has no stem or leaves. About nine months later, this bud produces a spectacular bloom that measures as much as 3 feet (1 meter) in diameter and 37 pounds (81 kilograms). It is the largest known flower in the world. The Rafflesia is not content to be just a parasite with a big, unattractive flower. It also produces one of the worst fragrances imaginable, the smell of rotting meat. The Rafflesia's goal is to attract flies with this fragrance in hopes that they will carry its pollen to a Rafflesia in the next forest.

Other fungi are parasites. Fungi reproduce by means of spores, which are usually single cells that have the ability to grow into a new organism.

Fungi prefer moist, dim environments, and they thrive on the shadowy forest floor. Some, like the marasmius, grow directly on the litter of plant matter, while others protrude from the trunks of trees. Another type of fungi, the mycorrhizae, live in the soil and surround the roots of most rain forest trees. They absorb energy from the tree and help the tree's roots absorb nutrients from the soil.

Lichens   Lichens are actually combinations of algae and fungi that live in cooperation. Fungi surround algae cells, and the algae obtain food for themselves and the fungi by means of photosynthesis. It is not known if fungi aid algal organisms, but it is believed that fungi may provide moisture for the algae.

Lichens often appear on rocks and other bare woodland surfaces. Some grow on the leaves of lowland trees, while others favor the cooler cloud forests and dangle from the limbs of trees. Lichens are common in all types of forests and seem able to survive most climatic conditions.

Green plants other than trees   Most green plants need several basic things to grow: sunlight, air, water, warmth, and nutrients. In the rain forest, water and warmth are abundant. However, nutrients such as nitrogen, phosphorus, and potassium, which are typically obtained from the soil, may not be in large supply. Light can be scarce because the thick forest canopy obscures the forest floor, and may block sunlight. The lack of seasons means that the canopy is in full leaf all year long. For this reason, most rain forest plants grow in the canopy of the forest. Those that do grow on the ground often have very large leaves that provide more surface area to be exposed to the scarce amount of light.

Common rain forest green plants   Common rain forest plants include rattans, pitcher plants, ferns, African violets, nasturtiums, Spanish mosses, orchids, lianas, urn plants, hibiscus shrubs, and bamboo.

Liana   Lianas are climbers found in rain forests throughout the world. Their roots can be large and tough, but they do not develop a thick trunk. Instead, they depend entirely on trees for support. Once they reach the canopy, they drape themselves among the branches and develop leaves, branches, flowers, and fruits. Often they send out "feeding," or aerial (AIR-ee-yuhl), roots that dangle in midair and absorb nutrients.

A common liana is the strangler fig, which begins as an epiphyte that sends long roots down to the ground. These roots grow branches that

Chocoholic Heaven

In the rain forests of Central and South America grows the cocoa, or cacao, tree. The seeds of this delightful tree produce chocolate. Initially, the roasted seeds were used by the Aztecs to make a hot drink flavored with vanilla and spices. In 1502, Italian explorer Christopher Columbus (1451–1506) brought the cocoa "beans" back to Spain, where the same drink was made but with sugar. Over the next 100 years, the use of cocoa spread to other parts of Europe and, by 1657, solid chocolate had been developed in France. Around 1700, the English improved hot chocolate by adding milk, and by 1850, manufacturing processes made sweet eating-chocolate possible. By 1876, its popularity had spread throughout the world.

Cocoa trees are raised on farms in Central and South America and portions of western Africa. In 1977, 1,653,000 tons (1,500,000 metric tons) of cocoa seeds were produced, and 20 percent of all the exports went to the United States. However, Americans consume only about 10 pounds (4.5 kilograms) of chocolate per person annually. The largest consumers are the Swiss, who eat 21 pounds (9.5 kilograms) per person each year.

Although the names are sometimes confused, the cocoa tree is not the source of the drug cocaine. Cocaine comes from the coca shrub, an unrelated plant.

surround the host tree, blocking the light. Eventually, the host dies and decays, leaving a hollow ring of stranglers.

Urn plant   Urn plants are epiphytes found in Central and South America that belong to the pineapple family. The plant's overlapping leaves form an urn, or cup, which collects rainwater and any dead plant matter that falls into it. Little hairs on the leaf surface absorb the water and dissolved minerals. Urn plants provide homes for many aquatic (water) insects and even frogs.

Hibiscus   Hibiscus shrubs grow on the ground around the edges of the forest where they can obtain light. In Africa, they quickly attain heights of up to 7 feet (2 meters). Their large, colorful flowers are bell shaped and may be scarlet, pink, yellow, or white.

Bamboo   Bamboo is a woody grass that can grow as tall as a tree. Dense forests of bamboo are found in Asia and Central Africa where plants may reach 130 feet (40 meters) in height. Bamboo tends to grow in thick, tightly packed clumps. Flowering occurs several years later, after which the plant dies. There are 480 species of bamboo. Its uses vary from food to instrument to paper.

Growing season   It is always growing season in the tropical rain forest. At any given time, at least one species will be flowering.

The viper snake has a prehensile tail, aiding its ability to move among the tree branches. They rarely venture down to the ground.
IMAGE COPYRIGHT SNOWLEOPARD1, 2007. USED UNDER LICENSE FROM SHUTTERSTOCK.COM.

Reproduction   Most green plants are seed plants, and most seed plants are flowering plants that reproduce by means of pollination. Pollination involves the transport of pollen from the stamen, the male part of the flower, to the pistil, the female part of the flower where seeds develop, by visiting animals or the wind. The seed's hard outer covering protects it while it waits for moisture and light to stimulate its growth.

Instead of pollination, some plants reproduce by means of rhizomes—long, rootlike stems that spread out below ground. These stems develop their own root systems and send up sprouts that develop into new plants.

Rain forest trees   Most trees have a single strong stem, or trunk. This single trunk gives them an advantage over smaller woody plants in that most of their growth is directed upward. Some large rain forest trees develop buttresses, winglike thickenings of the lower trunk that give tall trees extra support.

As a tree grows, its trunk is thickened with a new layer, or ring, of conducting tissues that carry water and nutrients from the roots to the branches. As the tree ages, the tissues from the center outward become hardened to produce a sturdy core. In cooler climates, the rings are formed seasonally and, when a tree is cut down, its age can be determined by how many rings are present. In the rain forest there are no seasons, so rings do not form regularly and cannot be used to accurately estimate a tree's age. Other methods are used to determine the age of rain forest trees, such as measuring the increase in the tree's circumference during a year and dividing that number into its total girth (size). Based on these calculations, one species in Malaysia may be 1,000 years old, a baby compared to the 4,900-year-old bristlecone pine of North America, but unusually old for the rain forest, where most trees only live 100 to 300 years.

Rain forest trees seldom exceed 200 feet (61 meters) in height, although the tualang of Malaysia has been recorded at 260 feet (80 meters). It is a common misconception that the world's tallest trees grow in the rain forests. The tallest trees grow in drier, more temperate climates.

The leaves of rain forest trees are often waxy and develop a "drip tip;" a long, narrow point that allows rain to run off easily.

Rainforests are full of thousands of different plant and animal species, many which may not have been discovered yet. IMAGE COPYRIGHT STEFFEN FOERSTER PHOTOGRAPHY, 2007. USED UNDER LICENSE FROM SHUTTERSTOCK.COM.

Common rain forest trees Common rain forest trees include black ebony, cinchona, mahogany, and mango.

Black ebony Found in the rain forest of Africa, the black ebony tree is almost white in color, but its heartwood is black and extremely hard. Valued for its heaviness, color, and durability, ebony heartwood is considered a precious wood. It has been commonly used to make the black chord keys on a piano.

Cinchona The cinchona (sing-KOH-nah) is native to Central and South America where it is found in lower montane rain forests. Its flowers grow in white, pink, and yellow clusters and its bark is an important source of several medicines, including quinine (KWY-nine), a treatment for malaria. During the 1800s, cinchona seeds were brought to Java in Indonesia, where the trees are still raised commercially.

Mahogany The term mahogany is applied to almost 200 species of trees with reddish-brown wood, winged seeds, and small, greenish-yellow flowers. The first trees given the name are native to the West Indies, but the most commercially important mahoganies come from Central and South America. They are used in fine furniture and paneling. African and Philippine mahogany are also economically important.

Mango The mango tree was originally discovered in East India. In its wild state, it produces a fibrous fruit that tastes like turpentine. In some species, this fruit is poisonous. Cultivated trees produce a delicious fruit that is enjoyed throughout the world. In the fifth century BC the mango was brought to Malaysia and eastern Asia. In India, the mango tree is sacred, believed to be a symbol of love and having the potential to grant wishes. The tree can reach heights of about 90 feet (27 meters) and is popular for its dense shade.

Growing season   Trees are woody perennial plants, which means they live more than one year, or growing season. When temperatures are warm year-round and rainfall is constant, as they are in the rain forest, trees become evergreen and grow almost continuously. Some rain forest trees shed their leaves periodically for a short time; however, this shedding is not simultaneous, even among trees of the same species.

Many rain forest trees do not bear fruit every year and thus do not regenerate readily. Seeds that lie on the forest floor may remain dormant (inactive) for many years. Those that sprout grow very slowly after the nutrients stored in the seeds are used up. If a tree falls and creates a gap in the canopy allowing sunlight to enter, these seedlings make up for lost time. They grow quickly toward the light and soon the gap is closed again.

Reproduction   In general, trees are divided into two groups according to how they bear their seeds. Gymnosperms produce seeds inside cones. Most conifers, like the pine, are gymnosperms. Angiosperms have flowers and produce their seeds inside fruits. Broad-leaved trees, such as maples, are usually angiosperms. Some species of gymnosperms are found in rain forests, but most rain forest trees are angiosperms.

The seedlings of some rain forest trees do not develop large amounts of chlorophyll, the substance in leaves that gives them their green color, until they reach light in the canopy. The leaves of these young trees are often red, blue, purple, or white instead of green.

Endangered species   Vast areas of rain forest have been destroyed by uncontrolled logging and clearing of the land for farms. From 2000 to 2005, tens of thousands of miles of rain forest in Brazil were lost due to deforestation. In more mountainous regions, such as Papua New Guinea, the land is less useful for farming and huge tracts of rain forest still remain untouched.

Many individual plants are threatened as the forest is destroyed. The African violet, for example, which is commonly cultivated as a house plant, is found in only a few forests in Tanzania, where it is rapidly disappearing.

Animal Life

Animal life in the rain forest is as diverse as its plant life because the warm temperatures and plentiful moisture aid survival. Most animals live in the trees, especially high in the canopy.

Microorganisms Microorganisms, like their name suggests, can not be seen without the aid of a microscope. Bacteria are microorganisms that are always present in forest soil. They help decompose dead plant and animal matter. They grow quickly in the warm, humid rain forest environment where they feed on the leaves, twigs, and other matter that falls from the canopy.

Invertebrates Animals without backbones are called invertebrates. They include simple animals such as insects and worms, and more complex animals such as the click beetle or the trapdoor spider. Certain groups of invertebrates, like mosquitoes, must spend part of their lives in water. In general, these types are not found in the trees, but in streams or in pools of rainwater. The humid rain forest is an ideal environment for many soft-bodied invertebrates, such as leeches, because there is little danger of drying out.

> ### The Web of Life: Biotic Potential
>
> The highest rate of reproduction under ideal conditions is a population's biotic potential. For most creatures, this potential is enormous. A single bacterium, for example, could set off a chain reaction of births that would cover our planet within thirty-six days.
>
> What keeps bacteria, as well as mosquitoes, frogs, and alligators reaching their full biotic potential? The answer is that, while creatures are determined to reproduce, life is not easy. Each life form is opposed by limiting factors that keep population growth in check. Cold temperatures, for example, kill mosquitoes. A shortage of mosquitoes starves frogs. A shortage of frogs keeps alligators hungry, preventing them from reproducing more quickly and taking over the world.

Common rain forest invertebrates In addition to bacteria, invertebrates remain the least known variety of life in the rain forest. Among those that have been identified are the Hercules beetle, the forest termite, the orchid bee, the Queen Alexandra's birdwing butterfly, the postman butterfly, the blue hunting wasp and the army ant.

Postman butterfly The postman butterfly lives in jungles of Central America. In its larval state as a spiny caterpillar, it is a voracious eater with a preference for the passionflower vine. The adult butterfly feeds on this vine's protein-rich pollen and nectar. The added protein gives it a longer lifespan than most butterflies, usually from six to nine months. The passionflower is a highly poisonous plant, so female butterflies lay their eggs on its youngest leaves, which contain less poison. As the larvae grow, they gradually absorb some of the poison and become immune to it.

The Web of Life: How Ants Help to Make Rain

Seventy-five percent of the rain that falls on the rain forests is recycled. It evaporates and creates clouds that protect the forest from too much sun. Many creatures help in this process. For example, the forests of the Amazon are home to vast armies of ants. Ants produce formic acid, which they use for defense and as a means of communication. They spray approximately 200,000 tons (181,400 metric tons) of formic acid into the air each year. This formic acid makes the rain that falls in the area slightly acidic and promotes the decay of dead wood. As the wood decays, bacteria are released into the atmosphere, where ice crystals tend to form around them. The ice crystals fall as rain, and the cycle continues.

Blue hunting wasp   The blue hunting wasp prefers dining on crickets, which it hunts by flying low over the forest floor. It grips a victim in its powerful jaws and then paralyzes it with its stinger. Female wasps drag the paralyzed victim into a burrow and then lay their eggs on it so their larvae have food when they hatch.

Army ant   Army ants are some of the most feared residents of the rain forest because they travel in huge colonies of up to 20 million individuals. The army's number demands that food be plentiful. The army usually empties an area of insects, small lizards, and snakes, leaving them no choice but to continue moving. They take shelter at night by linking themselves together with their strong legs beneath the fallen leaves and trees. When they are on the march, worker ants carry any developing larvae with them.

Food   Many invertebrates eat plants or decaying animal matter. The larvae of insects, such as caterpillars, are the primary leaf eaters. Weevils drill holes and lay their eggs in nuts, which their larvae use for food. Bees gather pollen and

The Postman butterfly is one of the pollinators of flowers in the rain forest. IMAGE COPYRIGHT TIM ZUROWSKI, 2007. USED UNDER LICENSE FROM SHUTTERSTOCK.COM.

nectar from flowers, as do butterflies and moths. The arachnids (spiders), which are carnivores (meat eaters), prey on insects and sometimes, if the spiders are big enough, small lizards, mice, and birds.

Reproduction   Most invertebrates have a four-part life cycle. The first stage of this cycle is spent as an egg. The eggshell is usually tough and resistant to long dry spells in tropical climates. After a rain, and during a period of plant growth, the egg hatches. The second stage is the larva, which may be divided into several stages between which there is a shedding of the outer skin as the larva increases in size. Larvae often spend their stage below ground where it is cooler and moister than on the surface. The pupal, or third stage, is spent hibernating within a casing, like a cocoon. In the fourth and final stage, the animal emerges from this casing as an adult.

The Spiny bush cricket is native to the rainforests of Central and South America. IMAGE COPYRIGHT DR. MORLEY READ, 2007. USED UNDER LICENSE FROM SHUTTERSTOCK.COM.

Amphibians   Amphibians are vertebrates (animals with a backbone) that usually spend part, if not most, of their lives in water. Frogs, toads, and salamanders are all amphibians. They live in significant numbers in rain forests where humid conditions are ideal. Amphibians are cold-blooded, which means their bodies are about the same temperature as their environment. In the rain forest they can be active year-round because the temperature is always warm.

Amphibians breathe through their skin, and only moist skin can absorb oxygen so they must usually remain close to a water source. Mating, egg-laying, and young-adulthood all take place in ponds, lakes, or pools of rainwater. When they mature, amphibians leave the pools for dry land where they feed on both plants and insects.

Common rain forest amphibians   Amphibians commonly found in rain forests include tree frogs and poison arrow frogs.

Tree frog   Eighty percent of frogs and toads live in tropical forests. Some, such as the fringe limb tree frog, may leap out of a tree to escape a predator. Webs of skin between their limbs act like parachutes, enabling them to glide from one branch to another. Other species of tree frogs have suction pads on their toes that secrete a sticky mucus, enabling the frog to cling to tree trunks and branches.

Tree frogs, such as this red-eyed tree frog, are common in the rain forest. Many tree frogs have suction cups on their feet to help them cling to smooth surfaces. IMAGE COPYRIGHT SNOWLEOPARD1, 2007. USED UNDER LICENSE FROM SHUTTERSTOCK.COM.

Poison arrow frog   Poison arrow frogs are usually brightly colored. This color warns potential predators that to bite the frog may mean death. One ounce of poison from the kokai frog can kill up to 100,000 average size humans. These frogs, abundant in South America, are used by the local people, who tip their arrows and darts in the poison (hence the frogs' name).

Food   Most adult amphibians are carnivorous, feeding on insects, slugs, and worms. Salamanders that live in the water suck their prey into their mouths. Those that live on land have long, sticky tongues to capture food. Salamander larvae are mostly herbivorous, feeding on vegetation. Frogs and toads feed on algae, plants, and insects such as mosquitoes.

Reproduction   Mating and egg-laying for amphibians must take place in water because male sperm are deposited in the water and must be able to swim to the eggs in order to penetrate them. Some amphibians lay their eggs in the cups of plants where water has collected and where insect larvae may grow. As the young develop into larvae and young adults, they often have gills for breathing. They, too, require a watery habitat.

Reptiles   Reptiles that live in rain forests include snakes and lizards. Since a reptile's body temperature changes with the temperature of the surrounding air, the warm, humid rain forest is a comfortable environment for them.

Many rain forest reptiles are capable of camouflage (KAH-mah-flahj), or protective coloration. Their skins are often patterned or colored to resemble the forest background in which they live, and they may be able to alter their coloration to a darker or lighter shade. Many reptiles living in the forest canopy have a prehensile (grasping) tail to help them climb and may prevent falls. Rain forest reptiles may have grasping claws to ensure a firm, steady hold as they climb through the trees.

Common rain forest reptiles Common rain forest snakes include the vine snake of West Africa, the bushmaster and fer-de-lance of Central and South America, the gaboon viper of Asia, and many tree snakes. Lizards include the Jesus Christ lizard of Central America, the crested water dragon of Asia, Parson's chameleon of Africa, and the Komodo dragon of Indonesia.

Chameleon The chameleon (kuh-MEEL-yuhn) lizard is an expert at camouflage. A resident of the canopy, it can change its coloration to resemble that of the leaves, and it may even tremble slightly to mimic leaves swaying in a breeze. Its feet and tail are perfect for grabbing hold of tree limbs, and its long, sticky tongue flicks out with amazing speed to catch the insects that make up its diet.

Tree snake Many species of tree snakes are found in rain forests. The emerald tree boa of the Amazon is not poisonous, and kills its prey with by squeezing it. Pythons, anacondas, and cobras are found in the rain forest. Pythons can be found in Africa and Australia, cobras in Africa and Asia, and the anaconda in South America.

Komodo dragon The Komodo dragon, the world's largest lizard, is found in Indonesia, especially on Komodo Island for which it is named. Komodo dragons can measure up to 10 feet (3 meters) in length and weigh up to 366 pounds (166 kilograms). Komodo dragons are carnivores and eat animals as large as buffalo or as small as geckos and other insects. They have long, sharp claws and jagged teeth that enable them to tear meat from their prey. In 1992, komodo dragons hatched at the Smithsonian Zoo, the first of their kind to ever have been bred outside of Indonesia.

Food A lizard's diet varies, depending upon the species. Some have long tongues with sticky tips for catching insects, while others eat small mammals and birds. The water they need is usually obtained from the food they eat.

All snakes are carnivores, and one good meal will last them for days or weeks. Some snakes kill their prey with venom (poison) injected through their fangs.

Reproduction Most reptiles lay eggs. Some females remain with the eggs, others bury them in a hole and abandon them leaving the young

The Emerald tree boa is one of many breeds of snake that thrive in the warmth and humidity of the rain forest.
IMAGE COPYRIGHT HOWARD SANDLER, 2007. USED UNDER LICENSE FROM SHUTTERSTOCK.COM.

Green Mansions

The South American rain forest provided the setting for *Green Mansions*, a novel by British author W. H. Hudson (1841–1922) that was written in 1904. It is a fantasy love story about Rima, a strange half-bird, half-human character who lives in the forest and cannot leave it. A statue of Rima was erected in 1925 in the bird sanctuary at Hyde Park in London, England.

to hatch by themselves. Snakes that live in the canopy often bear live young. They produce fewer babies, but the babies, being mobile, have a better chance of survival.

Birds   All rain forests have large bird populations. Most do not need the protection of camouflage, having only to compete with bats, and their feathers are usually brilliantly colored. Their songs vary from the scream of the eagle to the haunting warble of many smaller birds. Bower birds and pittas seem to be able to "throw" their voices like ventriloquists.

Rain forest birds have developed short, broad wings that do not require much room for flying because the vegetation is so dense. Some species, such as toucans and parrots, have feet adapted to climbing.

Common rain forest birds   Common birds of the rain forests include toucans, hummingbirds, birds of paradise, jacamars, eagles, parrots, and junglefowl.

Harpy eagle   Harpy eagles live high in the forest canopies of Central and South America, often sitting in one of the emergent trees where their sharp eyes can spot prey. Harpies do not soar high above, but move from tree to tree in short flights. Their large nests are built of sticks, leaves, and fur about 165 feet (50 meters) above the ground. Females produce two eggs. Harpies like to catch monkeys or sloths in their huge talons, and, if their victim attempts to cling to a tree, they are strong enough to wrench the victim free.

Parrot   Parrots are brightly colored birds with a loud, harsh call. There are 328 species of parrots, including the commonly known parakeets, macaws, and cockatoos. They tend to be social and roost in large groups. Their feet are strong enough to make it possible for them to hang upside down from a branch for long periods. Their nests are usually in holes in trees, where the female sits on the eggs, and the male brings her food. Preferring to eat seeds, parrots use their tongues to position a seed at the front of the strong, hooked beak and then crack the seed apart. Parrots, well know for their ability to mimic humans, are popular and desirable pets.

Junglefowl   The junglefowl of Asia is the ancestor of the chicken and, as such, has affected human life more than any other bird. Male junglefowl crow like roosters and have a red comb. In the wild, they are very aggressive; much of this behavior has been bred out of the domesticated species over the years. They live primarily on the forest floor where they feed on seeds, fruits, berries, and insects. They are good runners, but their flight is weak.

Food Rain forest birds may fly considerable distances in search of trees bearing fruit. Different birds seek food in different layers of the forest. Parrots, for example, hunt for seeds in the canopy and insects along the trunks of the trees, while pittas dig around on the ground for snails and ants.

Reproduction Birds reproduce by laying eggs, for which many species build a nest. Some, such as the macaw, prefer to use a hole in a tree. Females usually sit on the eggs until the young birds hatch. The female hornbill of Southeast Asia walls herself into a hole with mud and other materials and the male feeds her until the young can leave the nest. Both parents of most species usually feed their young until they are able to fly. Some young birds, such as the hoatzin of the Amazon region, are uniquely adapted to life in the forest. The hoatzin chicks have two claws on each wing, enabling them to climb through the branches.

Wild Blue and gold macaws live high in the trees in the rainforests of Mexico, Central and South America. IMAGE COPYRIGHT JANPRCHAL, 2007. USED UNDER LICENSE FROM SHUTTERSTOCK.COM.

Mammals Mammals of all kinds live in the rain forest, from the monkeys that swing from the tops of its trees to the shrews that scamper about the jungle floor.

Common rain forest mammals Rain forest mammals include monkeys, antelopes, coatis, bats, sloths, okapis, gorillas, and jaguars.

Bat Bats are the only mammals truly capable of flight, which they do primarily at night. During the day, they sleep hanging upside-down from branches or holes in trees. Some species are very social and roost in groups of 100 or more. At night they leave their roosts to seek food. A rain forest bat's diet is either insects or flower nectar and fruits. Many trees are dependent upon them as an aid in pollination.

The God of the Air

The quetzal, or resplendent trogon, of the mountain jungles of Central America has some of the most colorful plumage of any bird. To the Mayas and Aztecs, the quetzal's emerald and crimson feathers symbolized spring vegetation. They used its tail feathers in religious ceremonies and worshiped the bird as a god of the air. The beautiful quetzal is the national emblem of Guatemala, and its name is given to a unit of Guatemalan currency.

Sloth   Sloths are slow-moving creatures with large claws that spend their lives hanging upside-down from tree limbs in Central and South America. Adults are only about 2 feet (0.6 meters) long, and the claws by which they grip the trees measure about 3 inches (8 centimeters). Sloths are so well adapted to their life upside-down that even their hair grows that way, from stomach to spine. They are herbivores that almost never leave the trees because, on the ground, a sloth cannot stand or walk but rather drags itself around with its claws.

Okapi   The okapi, a short-necked relative of the giraffe that lives in the rain forest of eastern Congo, was not discovered by European explorers until 1901. It feeds on understory vegetation, such as shrubs and leaves. The coat is purplish brown, with black and white stripes on the upper legs and buttocks. Okapis are unusual in that females are larger than males.

Sloths eat leaves almost exclusively and rarely venture to the ground to move to a different tree. IMAGE COPYRIGHT ALVARO PANTOJA, 2007. USED UNDER LICENSE FROM SHUTTERSTOCK.COM.

Gorilla   Gorillas are found only in Africa, where there are three species: the western lowland gorilla, the mountain gorilla, and the eastern lowland gorilla. These types vary slightly in physical characteristics, such as color.

The gorilla is the world's largest living primate (group of animals including apes, monkeys, and humans). Males may be as tall as 6 feet (1.8 meters) and weigh between 300 and 400 pounds (135 and 180 kilograms). Gorillas live in groups and rarely change locations if they can help it. They spend most of their day, apart from a midday rest period, foraging for food, primarily nuts, berries, fruits, and leaves.

Jaguar   The largest cat in the western hemisphere, jaguars are found mostly in South America. A male jaguar may be 6 feet (1.8 meters) in length with a 2-foot (0.6-meter) -long tail, and may weigh more than 300 pounds (135 kilograms). It has a tan coat and is spotted like a leopard, though some are completely black or white. Jaguars feed on both small and large animals. They hunt mostly on the ground, but they are agile (skilled) climbers and swimmers.

Food   Small mammals, such as the fawn-footed melomys (a type of rodent), often eat plants and insects. Herbivores (plant-eaters), including the agouti, the paca, and the royal antelope, feed on leaf buds and fruit. Cats, such as the palm civet, the ocelot, and the servaline genet, are carnivores (meat-eaters). Many mammals, such as the orangutan, are omnivores, which means they eat both plant and animal foods.

Reproduction   The young of mammals develop inside the mother's body, where they are protected from predators. Mammals produce milk to feed their young and must remain nearby until the young can survive on their own.

World's Smallest Mammal

The world's smallest mammal is only 1 inch (2.5 centimeters) long, about the size of a bumblebee. It was first catalogued by researchers in 1974. It may have been overlooked until then because it only comes out at night and might have been mistaken for an extremely large mosquito. This little mammal is the Kitti's hog-nosed bat, and it lives in the rain forests of Thailand.

The Okapi live in the rainforests of central Africa. They were discovered by scientists in the 1900s.
IMAGE COPYRIGHT STEFFEN FOERSTER PHOTOGRAPHY, 2007. USED UNDER LICENSE FROM SHUTTERSTOCK.COM.

Gorillas are one of the many endangered species of the rain forest. IMAGE COPYRIGHT PICHUGIN DMITRY] 2007. USED UNDER LICENSE FROM SHUTTERSTOCK.COM.

Endangered species The list of endangered rain forest animals is long. Many species of parrots are endangered, especially in Central and South America, because they are sought by animal dealers to sell as pets or because their habitat is disappearing. An example is the Spix's macaw, found in Brazil. In 1999, only one bird remained alive in the wild. Concentrated efforts to bring the bird out of near extinction resulted in sixty birds living in captivity. Reintroducing them to the wild has been difficult. Biologists are attempting to send females into the wild in hopes that they will mate with the lone male.

Orangutans and gorillas are in danger because they require deep forest cover and their habitat is rapidly disappearing. For some, their only hope of preservation seems to be zoos and wildlife sanctuaries. Many rain forest cats are also endangered because they have been hunted extensively for their skins. Although many are now protected, their numbers are so low there may not be enough animals left for successful breeding.

Only five species of rhinoceros remain in the world. They are protected, but remain threatened by poachers who kill them for their horns, and by the loss of their habitat. Rhinos live in Africa, India, and Southeast Asia. Although their senses of smell and hearing are well developed, they can not see very well, making them susceptible to attack from both humans and forest predators.

Human Life

Humans are creatures of the forest. Until they learned to hunt, humans gathered their food and made their dwellings among the trees of the forests.

Impact of the rain forest on human life   Forests have an important impact on the environment as a whole. From the earliest times, forests have provided humans with food and shelter, a place to hide from predators, and many useful products.

Environmental cycles   Trees, soil, animals, and other plants all interact to create a balance in the environment from which humans benefit. This balance is maintained in what can be described as cycles.

The oxygen cycle   Plants and animals take in oxygen from the air and use it for their life processes. This oxygen must be replaced, or life on Earth could not continue. Animals breathe in (inhale) oxygen and breathe out (exhale) carbon dioxide. Trees convert this carbon dioxide into oxygen during photosynthesis, releasing the oxygen into the atmosphere through their leaves.

Rain Forest Cousins

Chimpanzees are great apes, the closest living relatives to the human species. They resemble humans in some important ways, including their use of tools. Some chimp populations in western Africa use stone and wooden hammers to break open nuts. In eastern Africa, chimps have been observed to stick plant stems into termite nests to drive the termites out so they can eat them, and during a heavy rain they use leafy tree branches as umbrellas.

The Sumatran orangutan is a critically endangered resident of the rain forest. IMAGE COPYRIGHT MICHAEL STEDEN, 2007. USED UNDER LICENSE FROM SHUTTERSTOCK.COM.

The carbon cycle   Carbon dioxide is necessary to life, although too much is harmful. During photosynthesis, trees pull carbon dioxide from the air. This helps maintain the oxygen/carbon dioxide balance in the atmosphere. When trees die, the carbon in their tissues is returned to the soil. Decaying trees become part of Earth's crust, and after millions of years, this carbon is converted into oil and natural gas.

The water cycle   The root systems and fallen leaves of trees help build an absorbent covering on the forest floor that allows rain water to trickle down into the soil to feed streams and groundwater. In this way, forests help conserve water and protect the soil from erosion caused by heavy rain. When forests are cut down, the soil washes away and flooding is more common. For example, heavy rainfall in Indonesia has caused severe flooding, in part due to tree removal. Trees take up some of this rain water through their roots and use it for their own life processes. Extra moisture is released through their leaves back into the atmosphere, where it helps to form clouds.

The nutrient cycle   Trees get the mineral nutrients they need from the soil. Dissolved minerals are absorbed from the soil by the tree's roots and are sent upward throughout the tree. These mineral nutrients are used by the tree much like humans take vitamins. When the tree dies, these nutrients, which are still contained within parts of the tree, decompose. They are then returned to the soil making them available for other plants and animals.

Food   Forests are the home of game animals, such as birds, that provide meat for hunters and their families. Forests also supply fruits, nuts, seeds, and berries, as well as vegetation for livestock. Vanilla, for example, is made from the seed pod of a type of orchid found in Central and South American rain forests; nutmeg and cloves come from Asian rain forests; coffee beans are native to Africa and are products of Central America along with cocoa beans; and starfruits grow in Asia. It is estimated that rain forests contain more than 3,000 species of edible fruits and vegetables, and only about 200 of these have been cultivated for commercial use. With correct management these species might yet provide more food varieties for both humans and animals.

Shelter   During prehistoric times, humans lived in the forest because it offered protection. Today trees can provide building materials. Trunks are cut into planks or used as poles, while fronds (branches) and grasses can be cut for thatch and used to make huts or roofs for wooden structures.

Economic values   Forests are important to the world economy. Many products used commercially, such as wood, medicines, tannins, dyes, oils, and resins (sap) are obtained from forests. Forest land is also important to the farm and tourist industries.

Wood   Trees produce one of two general types of wood, hardwood or softwood, based on the tree's cell wall structure. Hardwoods are usually produced by angiosperms, such as the mahogany tree, while most coniferous trees, such as pines, produce softwoods. These names can be confusing because some softwood trees, such as the yew, produce woods that are harder than many hardwoods. Some hardwoods, such as balsa, are softer than most softwoods.

Wood is used for fuel, building structures, and manufacturing other products, such as furniture and paper. Wood used for general construction is usually softwood. In order to conserve trees and reduce costs, some manufacturers have created engineered wood composed of particles of several types of wood combined with strong glues and preservatives. Engineered woods are very strong and can be used for many construction needs.

In an effort to preserve the natural forests, plantations are developed to supply the world's demands for wood. India, South America, and Africa combined have 102,000 square miles (264,178 square kilometers) of tree plantations.

Farmland   Although rain forest land is poor for farming, more and more of it is being cleared for that purpose. It supports crops for a few years, and then it is used for cattle pasture. When its nutrients are completely exhausted it is abandoned.

Medicines   Since the earliest times, plants have been used for their healing properties. It is estimated that at least 70 percent of cancer fighting plants are tropical plants from the rain forest. Quinine, from the cinchona tree, is used to fight malaria.

Coffee—The World's Most Valuable Agricultural Product

In the mid-1990s, imports of coffee beans to the United States totaled $1.5 billion each year. The United States is the world's largest importer, and the average American drinks about 27 gallons (102 liters) of coffee annually.

Coffee trees grow in montane forests, and coffee beans are the roasted seeds of the tree. The birthplace of coffee was probably Ethiopia, the tree's native environment. Its use was first developed by the Arabs, and it did not arrive in Europe until the sixteenth century where it was introduced as a medical potion. Coffee became popular as a beverage around 1652. Coffee plantations were soon established in Indonesia, the West Indies, and Brazil, and coffee cultivation became important to colonial economies. Latin America and Africa produce most of the world's coffee; Central and South America grow about 60 percent of the world's total production.

Deadly Traveler on the Kinshasa Highway

The Kinshasa Highway crosses central Africa, linking remote areas to airports in the large cities of Nairobi and Mombasa, both in Kenya. Any traveler starting from deep inside the rain forest who reaches one of those airports is within 24 hours of every other place on Earth. During the 1970s one traveler made such a journey, a journey of deadly consequence.

The traveler was HIV, the human immunodeficiency virus, and by 2007, only twenty-five years after its emergence from the rain forest, it had infected 40 million people worldwide. Already 25 million people have died of AIDS (acquired immune deficiency syndrome), which appears to be caused by HIV.

For millennia, potentially deadly bacteria and viruses have remained undisturbed in the rain forests. If humans caught them, villages were so remote and travel so slow that infected people seldom lived long enough to reach large populated areas where the disease could spread unchecked. As humans penetrate farther into the rain forest and destroy the natural balance that often keeps disease carriers under control, more and more problems are being created. The latest cause for concern is the hemorrhagic fever viruses, such as *Ebola zaire*, another rain forest resident, which is even more deadly than HIV.

Tannins, dyes, oils, and resins   Tannins are chemical substances found in the bark, roots, seeds, and leaves of many plants. It is used to cure leather, making it soft and supple. Dyes used to color fabrics can be obtained from the bark or leaves of such trees as the brazilwood. Palm oil and coconut oil are used as cooking ingredients. Resins, or saps, are used in paints and other products. Chicle is a resin from the sapodilla tree used in chewing gum, and natural rubber is made from the resin of the South American rubber tree.

Tourism   Rain forests have become popular with tourists who are interested in hunting, nature study, and environmental issues (ecotourism). Some tropical countries have found it economically desirable to set aside large tracts of forest for tourism.

Impact of human life on the rain forest   While forests have had a positive effect on human life, human life has had a mostly negative impact on forests. Nearly 60,000 square miles (155,399 square kilometers) of forest are cleared each year. Most of the forest loss has occurred in developing nations where wood is used for fuel and trees are cleared for farming. A large number of trees are also lost each year to commercial use.

Rain Forest Explorers

When Europeans first encountered the rain forest, they saw it from a position at its edge, where light could penetrate and foliage ran rampant. The forest soon gained a reputation for being impenetrable, and the first men who ventured into it had remarkable courage. Many were seekers of knowledge about the world; others had a hunger for riches that they believed they would find in the jungle.

At first, valuable goods were the primary target. The Portuguese penetrated Africa by means of its rivers but gained little knowledge of the interior. Few who entered it, such as Portuguese explorer Vallarte in 1448, came back alive. The first European to explore the rain forests of the Amazon in South America was Francisco de Orellana who went in search of cinnamon trees and gold in 1541.

In the 1700s, scientific curiosity became the primary motive for exploration. Between 1799 and 1803, Alexander Baron von Humboldt and Aime Bonpland explored thousands of miles of South America in order to study plant and animal life. During the nineteenth century, journalist Henry Morton Stanley (1841–1904) made a dangerous voyage the length of the Congo River in Africa. Prince Maximilian of Wied-Neuwied, Henry Walter Bates, Alfred Russel Wallace, Jules Crevaux, and, in the twentieth century, Theodore Roosevelt (1858–1919), Percy Fawcett, and Michael Rockefeller also undertook journeys into unexplored regions.

Use of plants and animals   Rain forests are disappearing at alarming rates each year, especially the montane forests. Much forestland is being lost as populations grow and want the land for farms and cattle pasture. Slash and burn agriculture is a routine procedure in which trees are cut and burned. When the land will no longer support crops, it is abandoned and additional forest is cut down somewhere else. As a result, many animals and plants are permanently losing their habitats.

Our ability to harvest trees for wood is greater than the forest's ability to regenerate. Mechanical harvesting with huge machines makes clear-cutting (cutting down every tree in an area) cheaper and more efficient than selecting only certain trees for harvesting. Replanting may not be done, or a fast-growing species may be replanted rather than the original species. The original species may never grow back. Clear-cutting endangers wildlife by destroying natural habitats.

Quality of the environment   Destruction of the forests does not mean just loss of their beauty and the products they provide. Water quality suffers because the trees are gone and rain no longer seeps into the soil. Instead, it runs off and underground water reserves are not

A Penan woman holds her child, deep within Sarawak's rainforest in Malaysian Borneo. The Penan are among the world's isolated communities. They hunt wild pigs and deer with spears and blow darts, and pick wild fruits from the lush bush.
AP IMAGES.

replaced. Topsoil is eroded away and often ends up in streams and rivers. If the quantity of soil is large enough, fish may die.

Air quality is also reduced by the destruction of forests. Trees not only put oxygen back into the air, but soot and dust floating in the air often collect on their leaves and are washed to the ground when it rains. When the trees are cut down, the dust and soot remain in the air as air pollution.

With the popularity of the automobile, carbon dioxide and other undesirable gases have built up in the atmosphere. Most scientists believe these gases are helping raise the temperature of Earth's climate (the greenhouse effect). Forests help remove carbon dioxide from the air, so cutting them down may be a factor in global warming. If Earth grows warmer, many species of plants and animals could become extinct.

Forest management Most rain forests grow in developing countries that need forest resources for economic reasons. Since 1945, more than 50 percent of the world's rain forests have been cut down and cleared away. Some countries have realized that they must use their resources wisely, and conservation efforts are under way. Malaysia and Uganda, for example, are making better use of trees that were formerly wasted, and replanting programs have begun in Gabon and Zambia.

Native peoples   Native people have been found in all the major rain forests of the world. They include the Yanomami of South America, the Asanti and Bambuti of Africa, the Andaman of Asia, the Aeta of the Phillipines, the Penan of Borneo, and the Aborigines of Australia.

In general, forest dwellers are hunter-gatherers who make few changes to the forest environment. Changes other people are making may destroy their way of life.

Yanomami   The Yanomami tribe, comprised of four smaller tribes, are scattered across Brazil and Venezuela. Only about 20,000 of them remain, as their population has decreased by over 10 percent in the last twenty years. The Yanomami live in villages as few as 40 residents to as many as 300 people. Trade and marriage keep the villages in contact with one another; sometimes peacefully and other times not. They seldom leave the forest, as it provides them with all they need to survive. Huts are built from timber and vine and food is either grown or hunted.

Asante   The Asante (sometimes referred to as Ashanti) of Africa make their home in Togo and the Ivory Coast. Like many native tribes, they began as forest dwellers. Loss of their home due to the clearing of the forest has caused some to move to nearby towns. Farming is their main livelihood. They export plantains, bananas, yams, and other staples to the local market.

Although mostly a peaceful tribe, the Asante have their share of war and strife. In the seventeenth century, firearm trade caused a power struggle in which the Asante prevailed. Then in the nineteenth century, they fought the British seeking an independent government from the Republic of Ghana. This battle was lost, but the Asante were allowed to live in peace, and they now number over 600,000 people.

Andaman   The Andaman tribe is the only tribe of natives that never learned to make fire. They waited for fire to naturally occur, such as lightning fires, and then were careful to preserve the fire as long as possible.

Located on the Andaman and Nicobar islands in the Bay of Bengal, they survive by hunting, and their diet consists of no plants or vegetables. As islanders, they seek out iron from nearby shipwrecks and use it to make their weapons.

The Food Web

The transfer of energy from organism to organism forms a series called a food chain. All the possible feeding relationships that exist in a biome make up its food web. In the rain forest, as elsewhere, the food web consists of producers, consumers, and decomposers. An analysis of the food web shows how energy is transferred within the biome.

Green plants are the primary producers in the forest. They produce organic materials from inorganic chemicals and outside sources of energy, primarily the Sun. Trees and other plants turn energy into plant matter.

Animals are consumers. Plant-eating animals, such as certain insects and mice, are the primary consumers in the rain forest food web. Secondary consumers, such as anteaters, eat the plant-eaters. Tertiary consumers are the predators, like owls and leopards. Some, such as orangutans and humans, are omnivores.

Decomposers feed on dead organic matter and include fungi and animals like the vulture. In the moist rain forest, bacteria aid decomposition. When leaves fall to the ground, bacteria feed on the leaves and speed up the decomposition process.

Spotlight on Rain Forests

Rain forests of South America   More than half of all the world's rain forests are located in South America. The South American region contains all three types—lowland, montane, and cloud forest—and is dominated by the great Amazon River and its tributaries. The forests of South America claim 2.7 million square miles (7 million square kilometers). The region on the northwest coast of Colombia is not well explored, in contrast to the much-explored forests of northern Brazil. Bordering Paraguay is the Mato Grosso forest. Cloud forests occur in the mountains of Venezuela, Brazil, Peru, and Guyana.

Annual rainfall is more than 236 inches (600 centimeters), which makes this one of the wettest places in the world.

An estimated 70,000 species of plants can be found in South American rain forests, which includes about 2,500 species of trees. Some trees, such as mahogany and rubber trees, have gained worldwide commercial importance.

No one knows how many species of insects live in the South American rain forest because so many remain to be identified or even discovered. Among those that have been catalogued include the malachite butterfly and the postman butterfly.

More than half of the world's species of birds make their home in the Amazon basin. A few are migratory but most live in the rain forest year-round. They have the short, broad wings of true jungle birds and range in size from the large-billed toucan to the tiny hummingbird.

Comparatively speaking, this area supports few mammals. They tend to be small and shy and include deer, pacas, agoutis, capybaras, anteaters, tapirs, jaguars, bats, and many species of monkeys, including spider monkeys, woolly monkeys, and capuchins.

In the 1970s, a network of highways was constructed through the Brazilian rain forest in an attempt to develop the land for farming and provide access to mineral and timber resources. This has proved

Rain forests of South America
Location: Brazil, Bolivia, Peru, Ecuador, Colombia, Venezuela, Guyana, Surinam, and French Guiana

disastrous. Rain forest soil is poor; after a few years farms are abandoned and more land is cleared. In addition, roads are often impassable from January to July, during the season of heaviest rain and severe flooding.

Rain forests of Central America   Before the sixteenth century, the small countries of Central America were covered by rain forest. Forests now cover only about 196,000 square miles (507,000 square kilometers) of land. Some forests are protected, such as the Monteverde Cloud Forest Reserve in Costa Rica.

Like the South American forests, the Central American forests are rich in diversity, claiming to have 43,000 to 48,000 different species; 20,000 to 25,000 of these are not found anywhere else in the world. Flowering plants include the hotlips, the brown violet-ear, and many species of orchids. Trees include the massive guanacaste and the Gunnera.

Paper wasps, red-kneed tarantulas, tiger moths, false-leaf katydids, golden beetles, termites, land snails, and sally lightfoot crabs are among the invertebrates that live here. Frogs and toads are exist in great numbers. While some depend on camouflage, others are brightly colored or patterned. Boa constrictors and iguanas are representative of reptiles. Bird life is extremely varied, and Panama supports more species than the whole of North America. Some birds are migratory, spending only the winter months here. Mammals include howler monkeys, tapirs, peccaries, deer, and jaguars.

For more than 3,000 years, Central America was home to the great Aztec and Maya civilizations. The Aztec were centered in Mexico and the Maya in Belize and Guatemala. They were primarily agricultural peoples and had developed methods of irrigating (watering) and fertilizing crops such as corn, beans, and squash. These civilizations also developed forms of writing, books, maps, astronomy, and a very accurate calendar. But by 1521, Hernán Cortés (1485–1547) of Spain had conquered the Aztecs, and by 1550 the Mayans were also overcome and their great civilizations destroyed.

Rain forests of Africa   The lowland wet forest lines the West African coast, from Senegal to the Congo. Montane forest is found in central south Africa. The total area is much smaller than covered by rain forest in South America or Southeast Asia. Here the forest measures about 780,000 square miles (2.02 million square kilometers), or one third of the continent.

The climate in Africa is warm and wet, and annual rainfall is greater than 60 inches (1.5 meters). This helps to support plant and animal life,

Rain forests of Central America
Location: Parts of Mexico, Panama, Costa Rica, Nicaragua, Honduras, Guatemala, and Belize

Rain forests of Africa
Location: Zaire, Congo, Cameroon, Gabon, Nigeria, Liberia, Sierra Leone, Ivory Coast

including the African mahogany, obeche, and ebony trees. African forests are also home to many flowering plants. One, called the "flame of the forest" or the "flamboyant tree" yields huge scarlet flowers. Animals commonly found in the African forest include gorillas, bats, monkeys, apes, and many others.

Rain forests of Madagascar   Rain forest in Madagascar covers the east side of the island along the coast. Rainfall is about 139 inches (353 centimeters) annually. The island is vulnerable to storms at sea and much damage occurs periodically.

Vegetation is dense and of the montane variety. Trees include traveler trees and palms. Ferns, lianas, and epiphytes dominate the understory.

Invertebrates include grasshoppers, termites, cockroaches, mosquitoes, moths, and butterflies. Reptiles include chameleons, geckos, and lizards. Birds are numerous and include guinea fowls, herons, flamingoes, and owls. At one time, Madagascar was part of Africa, and animal life in both places has similar origins. The current island broke away from Africa some 50 million years ago. As a result, the animals in Madagascar developed in isolation, creating a home to many unique species. These include many species of chameleon, the mesite, the lemur, the tenrec, and the fossa.

Rain forests of Southeast Asia   Tropical forests in Asia can be found in Bhutan, Myanmar, Bangladesh, India, the Malay peninsula, Indonesia, and the Philippines. Together, it covers 864,000 square miles (12.2 million square kilometers) of land. Within that, 566,000 square miles (1.47 million square kilometers) is covered in lowland wet forest.

In terms of tree species, the rain forests of Asia differ from others in that they support coniferous trees. The dipterocarps is a large family of hardwoods bearing winged fruits. It dominates these forests. Examples of dipterocarps include the *Shorea* and *Dipterocarpus* species. Other trees include kapoks, palms, and even pines. Smaller plants include mosses, ferns, rattans, ginger, orchids, and ant plants.

Asian forests are rich in fruit, such as breadfruit, jackfruit, and durian. Jackfruit is one of the largest fruits in the world, with some species weighing up to 55 pounds (25 kilograms). Durian has a

Rain forests of Madagascar
Location: Madagascar Island, off the east coast of Africa

Rain forests of Southeast Asia
Location: Malayasia, Indonesia, the Philippines, Thailand, Cambodia, Laos, and Vietnam

very distinct, unpleasant odor, but is popular among the natives of Asia.

Mammals are well represented and include rats, squirrels, tigers, elephants, rhinoceroses, tapirs, wild pigs, leopards, deer, antelopes, marbled cats, and many species of monkeys and their relatives.

Rain forests of New Guinea   The eastern half of New Guinea, Papau New Guinea, is an independent country, while the western half, Irian Jaya, belongs to Indonesia. As a whole, the island boasts that 77 percent of its lands are mostly lowland wet forests. Ranges of mountains run through the center of the island, the tallest peak is 16,535 feet (5,040 meters) high. Much of this region is unexplored. Steep gorges and rolling valleys punctuate the mountains.

Rainfall is heavy and, on the north coast, totals about 100 inches (254 centimeters) annually. Lowland temperatures remain at about 80°F (27°C) throughout the year.

The forest is rich with kamamere, lancewood, New Guinea basewood, and walnut. Moretan bay pines and klinki pines can be found at higher altitudes in the montane forest. The klinki pine is the tallest tree in all of the tropical forests in the world, measuring 292 feet (89 meters) tall.

Constant rain removes nutrients from the soil, and most inland soils are poor. In regions around volcanoes soil quality improves. In the montane forests, dead vegetation several feet thick often covers the ground.

Animal life resembles that found in Australia, especially the mammals. Ants, cockroaches, sand flies, butterflies, mosquitoes, and giant snails are representative of invertebrates, and the island is home to many snakes, most of them poisonous. Other reptiles include lizards, tortoises, and crocodiles. Birds are similar to those found in Malaysia and Australia; the cassowary and the bird of paradise are examples. Many mammals are found here, all of which are marsupials (mammals that carry their young in a pouch) except for the spiny anteater, the bat, and non-native rodents. An example of a marsupial is the tree kangaroo.

Native peoples depend upon agriculture, and the rain forest is gradually being cut down to accommodate farms. Crocodile farming is common for the sale of their skins.

Rain forests of New Guinea
Location: Pacific Ocean, north of Australia

For More Information

BOOKS

Allaby, Michael. *Biomes of the Earth: Temperate Forests.* New York: Chelsea House, 2006.

Allaby, Michael. *Biomes of the Earth: Tropical Forests.* New York: Chelsea House, 2006.

Fisher, William H. *Rain Forest Exchanges: Industry and Community on an Amazonian Frontier.* Washington DC: Smithsonian Institution Press, 2000.

Grzimek, Bernhard. *Grizmek's Animal Encyclopedia*, 2nd edition. Volume 7. *Reptiles,* edited by Michael Hutchins, James B. Murphy, and Neil Schlager. Farmington Hills, MI: Gale Group, 2003.

Marent, Thomas, and Ben Morgan. *Rainforest.* New York: DK Publishing, 2006.

Primack, Richard B., and Richard Corlett. *Tropical Rain Forests: An Ecological and Biogeographical Comparison.* Hoboken, NJ: Wiley-Blackwell, 2005.

Vandermeer, John H., and Ivette Perfecto *Breakfast Of Biodiversity: The Political Ecology of Rain Forest Destruction.* Oakland, CA: Food First Books, 2005.

PERIODICALS

Coghlan, Andy. "Earth Suffers as We Gobble Up Resources." *New Scientist.* 195. 2611 July 7, 2007: 15.

Darack, Ed. "The Hoh Rainforest." *Weatherwise.* 58. 6 Nov-Dec 2005: 20.

Gunnard, Jessie, Andrew Wier and Lynn Margulis. "Mycological Maestros: in the Ecuadorean Rainforest, a 'Missing Link' in the Evolution of Termite Agriculture?" *Natural History.* 112. 4 May 2003: 22.

Stone, Roger D. "Tomorrow's Amazonia: as Farming, Ranching, and Logging Shrink the Globe's Great Rainforest, the Planet Heats Up." *The American Prospect.* 18. 9 September 2007: A2.

ORGANIZATIONS

African Wildlife Foundation, 1400 16th Street NW, Suite 120, Washington, DC 20036, Phone: 202-939-3333; Fax: 202-939-3332, Internet: http://www.awf.org.

Environmental Defense Fund, 257 Park Ave. South, New York, NY 10010, Phone: 212-505-2100; Fax: 212-505-2375, Internet: http://www.edf.org.

Environmental Protection Agency, 401 M Street, SW, Washington, DC 20460, Phone: 202-260-2090, Internet: http://www.epa.gov.

Friends of the Earth, 1717 Massachusetts Ave. NW, 300, Washington, DC 20036, Phone: 877-843-8687; Fax: 202-783-0444, Internet: http://www.foe.org.

Greenpeace USA, 702 H Street NW, Washington, DC 20001, Phone: 202-462-1177, Internet: http://www.greenpeace.org.

Rainforest Alliance, 665 Broadway, Suite 500, New York, NY 10012, Phone: 888-MY-EARTH; Fax: 212-677-1900, Internet: http://www. rainforest-alliance.org.

Sierra Club, 85 2nd Street, 2nd fl., San Francisco, CA 94105, Phone: 415-977-5500; Fax: 415-977-5799, Internet: http://www.sierraclub.org.

The Wilderness Society, 1615 M st. NW, Washington, DC 20036, Phone: 800-the-wild, Internet: http://www.wilderness.org.

World Wildlife Fund, 1250 24th Street NW, Washington, DC 20090, Internet: http://www.wwf.org.

WEB SITES

National Geographic Magazine. http://www.nationalgeographic.com (accessed August 22, 2007).

National Park Service. http://www.nps.gov (accessed August 22, 2007).

Nature Conservancy. http://www.tnc.org (accessed August 22, 2007).

Scientific American Magazine. http://www.sciam.com (accessed August 22, 2007).

The World Conservation Union. http://www.iucn.org (accessed August 22, 2007).

World Wildlife Fund. http://www.wwf.org (accessed August 22, 2007).

Where to Learn More

Books

Allaby, Michael. *Biomes of the Earth: Deserts.* New York: Chelsea House, 2006.

Allaby, Michael. *Biomes of the Earth: Grasslands.* New York: Chelsea House, 2006.

Allaby, Michael. *Biomes of the Earth: Temperate Forests.* New York: Chelsea House, 2006.

Allaby, Michael. *Biomes of the Earth: Tropical Forests.* New York: Chelsea House, 2006.

Allaby, Michael. *Temperate Forests.* New York: Facts on File, 2007.

Angelier, Eugene, and James Munnick. *Ecology of Streams and Rivers.* Enfield, New Hampshire: Science Publishers, 2003.

Ballesta, Laurent. *Planet Ocean: Voyage to the Heart of the Marine Realm.* National Geographic, 2007.

Batzer, Darold P., and Rebecca R. Sharitz. *Ecology of Freshwater and Estuarine Wetlands.* Berkeley: University of California Press, 2007.

Braun, E. Lucy. *Deciduous Forests of Eastern North America.* Caldwell, NJ: Blackburn Press, 2001.

Brockman, C. Frank. *Trees of North America: A Guide to Field Identification,* Revised and Updated. New York: St. Martin's Press, 2001.

Burie, David, and Don E. Wilson, eds. *Animal.* New York: Smithsonian Institute, 2001.

Carson, Rachel L. *The Sea Around Us*, Rev. ed. New York: Chelsea House, 2006.

Cox, C.B., and P.D. Moore. *Biogeography and Ecological and Evolutionary Approach*, 7th ed. Oxford: Blackwell Publishing, 2005.

Cox, Donald D. *A Naturalist's Guide to Seashore Plants: An Ecology for Eastern North America.* Syracuse, NY: Syracuse University Press, 2003.

Day, Trevor. *Biomes of the Earth: Lakes and Rivers.* New York: Chelsea House, 2006.

Day, Trevor. *Biomes of the Earth: Taiga.* New York: Chelsea House, 2006.

Fisher, William H. *Rain Forest Exchanges: Industry and Community on an Amazonian Frontier.* Washington DC: Smithsonian Institution Press, 2000.

Fowler, Theda Braddock. *Wetlands: an Introduction to Ecology, the Law, and Permitting.* Lanham, MD: Government Institutes, 2007.

Gaston, K.J., and J.I. Spicer. *Biodiversity: An Introduction*, 2nd ed. Oxford: Blackwell Publishing, 2004.

Gleick, Peter H., et al. *The World's Water 2004-2005: The Biennial Report on Freshwater Sources.* Washington DC: Island Press, 2004.

Gloss, Gerry, Barbara Downes, and Andrew Boulton. *Freshwater Ecology: A Scientific Introduction.* Malden, MA: Blackwell Publishing, 2004.

Grzimek, Bernhard. *Grizmek's Animal Encyclopedia*, 2nd ed. Vol 7. *Reptiles,* edited by Michael Hutchins, James B. Murphy, and Neil Schlager. Farmington Hills, MI: Gale Group, 2003.

Hancock, Paul L., and Brian J. Skinner, eds. *The Oxford Companion to the Earth.* New York: Oxford University Press, 2000.

Harris, Vernon. *Sessile Animals of the Sea Shore.* New York: Chapman and Hall, 2007.

Hauer, Richard, and Gary A. Lamberti. *Methods in Stream Ecology*, 2nd ed. San Diego: Academic Press/Elsevier, 2007.

Hodgson, Wendy C. *Food Plants of the Sonoran Desert.* Tucson: University of Arizona Press, 2001.

Houghton, J. *Global Warming: The Complete Briefing.* 3rd ed. Cambridge: Cambridge University Press, 2004.

Humphreys, L.R. *The Evolving Science of Grassland Improvement.* Cambridge, UK: Cambridge University Press, 2007.

Hurtig, Jennifer. Deciduous Forests. New York: Weigl Publishers, 2006.

Irish, Mary. *Gardening in the Desert: A Guide to Plant Selection & Care.* Tucson: University of Arizona Press, 2000.

Jacke, Dave. *Edible Forest Gardens.* White River Junction, VT: Chelsea Green Publishing Co., 2005.

Johansson, Philip. The Temperate Forest: A Web of Life. Berkely Heights, NJ: Enslow Publishers, 2004.

Johnson, Mark. *The Ultimate Desert Handbook: A Manual for Desert Hikers, Campers and Travelers.* Camden, Maine: Ragged Mountain Press/McGraw Hill, 2003.

Jose, Shibu, Eric J. Jokela, and Deborah L. Miller *The Longleaf Pine Ecosystem: Ecology, Silviculture, and Restoration.* New York: Springer, 2006.

Luhr, James F., ed. *Earth*. New York: Dorling Kindersley in association with The Smithsonian Institute, 2003.

Marent, Thomas, and Ben Morgan. *Rainforest*. New York: DK Publishing, 2006.

McGonigal, D., and L. Woodworth. *Antarctica: The Complete Story*. London: Frances Lincoln, 2003.

Miller, James H., and Karl V. Miller. *Forest Plants Of The Southeast And Their Wildlife Uses*. Athens: University of Georgia Press, 2005.

Mitsch, William J., and James G. Gosselink. *Wetlands*, 4th ed. Hoboken, NJ: Wiley: 2007.

Moore, Peter D. *Biomes of the Earth: Tundra*. New York: Chelsea House, 2006.

Moore, Peter D. *Wetlands*. New York: Chelsea House, 2006.

Moul, Francis. *The National Grasslands: A Guide to America's Undiscovered Treasures*. Lincoln: University of Nebraska Press, 2006.

National Oceanic and Atmospheric Administration *Hidden Depths: Atlas of the Oceans*. New York: HarperCollins, 2007.

Ocean. New York: DK Publishing, 2006.

O'Shea, Mark, and Tim Halliday. *Smithsonian Handbooks: Reptiles and Amphibians*. New York: Dorling Kindersley, 2001.

Oliver, John E., and John J. Hidore. *Climatology: An Atmosperic Science*, 2nd ed. Upper Saddle River, NJ: Prentice Hall, 2004.

Peschak, Thomas P. *Wild Seas, Secret Shores of Africa*. Cape Town: Struik Publishers, 2008.

Postel, Sandra, and Brian Richter *Rivers for Life: Managing Water For People And Nature*. Washington, DC: Island Press, 2003.

Preston-Mafham, Ken, and Rod Preston-Mafham. *Seashore*. New York: HarperCollins Publishers, 2004.

Primack, Richard B., and Richard Corlett. *Tropical Rain Forests: An Ecological and Biogeographical Comparison*. Hoboken, NJ: Wiley-Blackwell, 2005.

Romashko, Sandra D. *Birds of the Water, Sea, and Shore*. Lakeville, MN: Windward Publishing, 2001.

Scientific American, ed. *Oceans: A Scientific American Reader*. Chicago: University Of Chicago Press: 2007.

Scrace, Carolyn. *Life in the Wetlands*. New York: Children's Press, 2005.

Shea, John F., et al. *Wetlands, Buffer Zones and Riverfront Areas: Wildlife Habitat and Endangered Species*. Boston: MCLE, 2004.

Sowell, John B. *Desert Ecology*. Salt Lake City: University of Utah Press, 2001.

Sverdrup Keith A., and Virginia Armbrust. *An Introduction to the World's Oceans*. Boston: McGraw-Hill Science, 2006.

Swarts, Frederick A. *The Pantanal: Understanding and Preserving the World's Largest Wetland*. St. Paul, MN: Paragon House Publishers, 2000.

Vandermeer, John H., and Ivette Perfecto *Breakfast Of Biodiversity: The Political Ecology of Rain Forest Destruction.* Oakland, CA: Food First Books, 2005.

Vogel, Carole Garbuny. *Shifting Shores (The Restless Sea).* London: Franklin Watts, 2003.

Voshell, J. Reese, Jr. *A Guide to Common Freshwater Invertebrates of North America.* Blacksburg, VA: McDonald and Woodward Publishing Company, 2002.

Woodward, Susan L. *Biomes of Earth: Terrestrial, Aquatic, and Human Dominated.* Westport, CT: Greenwood Press, 2003.

Worldwatch Institute, ed. *Vital Signs 2003: The Trends That Are Shaping Our Future.* New York: W.W. Norton, 2003.

Yahner, Richard H. *Eastern Deciduous Forest, Second Edition: Ecology and Wildlife Conservation.* Minneapolis, MN: University of Minnesota Press, 2000.

Zabel, Cynthia, and Robert G. Anthony. *Mammal Community Dynamics: Management and Conservation in the Coniferous Forests of Western North America.* New York: Cambridge University Press, 2003.

Periodicals

"Arctic Sea Ice Reaches Record Low." *Weatherwise.* 60. 6 Nov-Dec 2007: 11.

Bischof, Barbie. "Who's Watching Whom?" *Natural History.* 116. 10 December 2007: 72.

Bright, Chris, and Ashley Mattoon. "The Restoration of a Hotspot Begins." *World Watch* 14.6 Nov-Dec 2001: p. 8–9.

Carwardine, Mark. "So Long, and Thanks for All the Fish: If the Yangtze River Dolphin isn't Quite Extinct Yet, It Soon Will Be." *New Scientist.* 195. 2621 September 15, 2007: 50.

Coghlan, Andy. "Earth Suffers as We Gobble Up Resources." *New Scientist.* 195. 2611 July 7, 2007: 15.

Cunningham, A. "Going Native: Diverse Grassland Plants Edge Out Crops as Biofuel." *Science News.* 170. 24 December 9, 2006: 372.

Darack, Ed. "Death Valley Springs Alive." *Weatherwise.* 58. 4 July-August 2005: 42.

Darack, Ed. "The Hoh Rainforest." *Weatherwise.* 58. 6 Nov-Dec 2005: 20.

De Silva, José María Cardoso, and John M. Bates. "Biogeographic Patterns and Conservation in the South American Cerrado: A tropical Savanna Hotspot." *BioScience* 52.3 March 2002: p225.

El-Bagouri, Ismail H.M. "Interaction of Climate Change and Land Degradation: the Experience in the Arab Region." *UN Chronicle.* 44. 2 June 2007: 50.

Ehrhardt, Cheri M. "An Amphibious Assault." *Endangered Species Bulletin.* 28. 1 January-February 2003: 14.

Flicker, John. "Audubon view: Grassland Protection." *Audubon.* 107. 3 May-June 2005: 6.

Greer, Carrie A. "Your Local Desert Food and Drugstore." *Skipping Stones.* 20. 2 March-April 2008: 34.

Gunnard, Jessie, Andrew Wier and Lynn Margulis. "Mycological Maestros: in the Ecuadorean Rainforest, a 'Missing Link' in the Evolution of Termite Agriculture?" *Natural History.* 112. 4 May 2003: 22.

Haedrich, Richard L. "Deep Trouble: Fishermen Have Been Casting Their Nets into the Deep Sea After Exhausting Shallow-water Stocks. But Adaptations to Deepwater Living Make the Fishes There Particularly Vulnerable to Overfishing—and Many are Now Endangered." *Natural History.* 116. 8 October 2007: 28.

Hansen, Andrew J., et al. "Global Change in Forests: Responses of Species, Communities, and Biomes." *BioScience.* 51. 9 September 2001: 765.

Johnson, Dan. "Wetlands: Going, Goings … Gone?" *The Futurist.* 35. 5 September 2001: 6.

"Katrina Descends." *Weatherwise.* 58. 6 November-December 2005: 10.

Kessler, Rebecca. "Spring Back." *Natural History* 115.9 November 2006: p 18.

Kloor, Keith. "Fire (in the Sky): In Less than an Hour, Flames had Reduced Nearly 8,000 acres of Grasslands to Smoldering Stubble and Ash." *Audubon.* 105. 3 September 2003: 74.

Lancaster, Pat. "The Oman experience." *The Middle East.* 384 December 2007: 55.

Levy, Sharon. *New Scientist: Last Days of the Locust.* February 21, 2004, p. 48–49.

Marshall, Laurence A. "Sacred Sea: A Journey to Lake Baikal." *Natural History.* 116. 9 November 2007: 69.

Maynard, Barbara. "Fire in Ice: Natural Gas Locked Up in Methane Hydrates Could Be the World's Next Great Energy Source—If Engineers Can Figure Out How to Extract it Safely." *Popular Mechanics.* 183. 4 April 2006: 40.

Mohlenbrock, Robert H. "Wetted Bliss: in a Louisiana Refuge, Different Degrees of Moisture Create Distinctive Woods." *Natural History.* 117. 2 March 2008: 62.

Morris, John. "Will the Forest Survive?" *Wide World.* 13. 1 September 2001: 15.

Payette, Serge, Marie-Josee Fortin, and Isabelle Gamache. "The Subarctic Forest—Tundra: The Structure of a Biome in a Changing Climate." *BioScience.* 51. 9 September 2001: 709.

Pennisi, Elizabeth. "Neither Cold Nor Snow Stops Tundra Fungi." *Science.* 301. 5,638 September 5, 2003: 1,307.

Petty, Megan E. "The Colorado River's Dry Past." *Weatherwise.* 60. 4 (July-August 2007): 11.

Pollard, Simon D, and Robert R. Jackson. "Vampire Slayers of Lake Victoria: African Spiders get the Jump on Blood-filled Mosquitoes. (*Evarcha culicivora*)." *Natural History.* 116. 8 October 2007: 34.

Shefrin, Russell. "An Optimistic Look at Falling Leaves." *New York State Conservationist*. 62.2 October 2007: p32.

"Siberia Sees the Wood from the Trees." *Geographical*. 73. 1 January 2001: 10.

Springer, Craig. "Leading-edge Science for Imperiled, Bonytail." *Endangered Species Bulletin*. 27. 2 March-June 2002: 27.

Springer, Craig. "The Return of a Lake-dwelling Giant." *Endangered Species Bulletin*. 32. 1 February 2007: 10.

Sterling, Eleanor J, and Merry D. Camhi. "Sold Down the River: Dried Up, Dammed, Polluted, Overfished—Freshwater Habitats Around the World are Becoming Less and Less Hospitable to Wildlife." *Natural History*. 116. 9 November 2007: 40.

Stone, Roger D. "Tomorrow's Amazonia: as Farming, Ranching, and Logging Shrink the Globe's Great Rainforest, the Planet Heats Up." *The American Prospect*. 18. 9 September 2007: A2.

Tucker, Patrick. "Growth in Ocean-current Power Foreseen: Florida Team Seeks to Harness Gulf Stream." *The Futurist*. 41. 2 March-April 2007: 8.

Wagner, Cynthia G. "Battles On the Beaches." *The Futurist*. 35. 6 November 2001: 68.

Weir, Kirsten L. "Don't Tread on It." *Natural History*. 110. 9 November 2001: 37.

Yalden, Derek. "Managing moorland." *Biological Sciences Review*. 18. 2 November 2005: 14.

Organizations

African Wildlife Foundation, 1400 16th St. NW, Suite 120, Washington, DC 20036. Phone: 202-939-3333, Fax: 202-939-3332, Internet: http://www.awf.org.

American Cetacean Society, PO Box 1391, San Pedro, CA 90733. Internet: http://www.acsonline.org.

American Littoral Society, Sandy Hook, Highlands, NJ 07732. Phone: 732-291-0055, Internet: http://www.alsnyc.org.

American Oceans Campaign, 2501 M St., NW Suite 300, Washington DC 20037-1311. Phone: 202-833-3900. Fax: 202-833-2070, Internet: http://www.oceana.org.

American Rivers, 1101 14th St. NW, Suite 1400, Washington, DC 20005. Phone: 202-347-7550, Fax: 202-347-9240, Internet: http://www.americanrivers.org.

Canadian Lakes Loon Survey, PO Box 160, Port Rowan, ON, Canada N0E 1M0. Internet: http://www.bsc-eoc.org/cllsmain.html.

Center for Environmental Education, Center for Marine Conservation, 1725 De Sales St. NW, Suite 500, Washington, DC 20036.

Center for Marine Conservation, 1725 DeSales St., NW, Suite 600, Washington, DC 20036. Phone: 202-429-5609, Fax: 202-872-0619, Internet: http://www.cmc-ocean.org.

Chihuahuan Desert Research Institute, PO Box 905, Fort Davis, TX 79734. Phone: 432-364-2499, Fax: 432-364-2686, Internet: http://www.cdri.org.

Coast Alliance, PO Box 505, Sandy Hook, Highlands, NJ 07732. Phone: 732-291-0055, Internet: http://www.coastalliance.org.

Desert Protective Council, Inc., PO Box 3635, San Diego, CA 92163. Phone: 619-342-5524, Internet: http://www.dpcinc.org/environ_issues.shtml.

Envirolink, PO Box 8102, Pittsburgh, PA 15217. Internet: http://www.envirolink.org.

Environmental Defense Fund, 257 Park Ave. South, New York, NY 10010. Telephone: 212-505-2100, Fax: 212-505-2375, Internet: http://www.edf.org.

Environmental Network, 4618 Henry St., Pittsburgh, PA 15213. Internet: www.environlink.org

Environmental Protection Agency, 401 M St., SW, Washington DC 20460. Telephone:202-260-2090, Internet: http://www.epa.gov.

The Freshwater Society, 2500 Shadywood Rd., Navarre, MN 55331. Phone: 952-471-9773, Fax: 952-471-7685, Internet: http://www.freshwater.org.

Friends of the Earth, 1717 Massachusetts Ave. NW 300, Washington, DC 20036. Telephone:877-843-8687, Fax:202-783-0444, Internet: http://www.foe.org.

Global ReLeaf, American Forests, PO Box 2000, Washington, DC 20013. Telephone: 202-737-1944, Internet: http://www.amfor.org.

Global Rivers Environmental Education Network (GREEN), 2120 W. 33rd Avenue, Denver, CO 80211. Internet: http://www.earthforce.org/green.

Greenpeace USA, 702 H St. NW, Washington, DC 20001. Telephone:202-462-1177, Internet: http://www.greenpeace.org.

International Joint Commission, 1250 23rd St. NW, Suite 100, Washington, DC 20440. Phone: 202-736-9024, Fax: 202-467-0746, Internet: http://www.ijc.org.

Izaak Walton League of America, 707 Conservation Ln., Gaithersburg, MD 20878. Telephone: 301-548-0150; Internet: http://www.iwla.org

National Wetlands Conservation Project, The Nature Conservancy, 1800 N Kent St., Suite 800, Arlington, VA 22209. Phone: 800-628-6860, Internet: http://www.tnc.org.

Nature Conservancy, Worldwide Office, 4245 North Fairfax Dr., Arlington, VA 22203-1606. Phone: 800-628-6860, Internet: http://www.nature.org.

North American Lake Management Society, PO Box 5443, Madison, WI 53705-0443. Phone: 608-233-2836, Fax: 608-233-3186, Internet: http://www.nalms.org.

Olympic Coast Alliance, PO Box 573 Olympia, WA 98501. Phone: 360-705-1549, Internet: http://www.olympiccoast.org.

Project Wet, 1001 West Oak, Suite 210, Bozeman, MT 59717. Phone: 866-337-5486, Fax: 406-522-0394, Internet: http://projectwet.org.

Rainforest Alliance, 665 Broadway, Suite 500, New York, NY 10012. Phone: 888-MY-EARTH, Fax: 212-677-1900, Internet: http://www.rainforest-alliance.org.

Sierra Club, 85 St., 2nd Fl., San Francisco, CA 94105. Telephone: 415-977-5500, Fax: 415-977-5799, Internet: http://www.sierraclub.org.

The Wilderness Society, 1615 M St. NW, Washington, DC 20036. Telephone: 800-the-wild, Internet: http://www.wilderness.org.

World Meteorological Organization, 7-bis, avenue de la Paix, Case Postale No. 2300 CH-1211, PO Box 2300, Geneva 2, Switzerland. Phone: 41 22 7308111, Fax: 41 22 7308181, Internet: http://www.wmo.ch.

World Wildlife Fund, 1250 24th St. NW, Washington, DC 20090. Internet: http://www.wwf.org.

Web sites

"America's Wetlands." Environmental Protection Agency. http://www.epa.gov/OWOW/wetlands/vital/what.html (accessed July 13, 2007).

Arctic Studies Center.http://www.nmnh.si.edu/arctic (accessed August 9, 2007).

Blue Planet Biomes. http://www.blueplanetbiomes.org (accessed September 14, 2007).

CBC News Indepth: Oceans. http://www.cbc.ca/news/background/oceans/part2.html (accessed September 1, 2007).

Discover Magazine. http://www.discovermagazine.com (accessed September 12, 2007).

Distribution of Land and Water on the Planet. http://www.oceansatlas.com/unatlas/about/physicalandchemicalproperties/background/seemore1.html (accessed September 1, 2007).

Envirolink. http://www.envirolink.org (accessed September 14, 2007).

FAO Fisheries Department. http://www.fao.org/fi (accessed September 5, 2007).

"The Grassland Biome." University of California Museum of Paleontology. http://www.ucmp.berkeley.edu/exhibits/biomes/grasslands.php (accessed September 14, 2007).

Journey North Project. http://www.learner.org/jnorth (accessed August 14, 2007).

Long Term Ecological Research Network. http://lternet.edu (accessed August 9, 2007).

Monterey Bay Aquarium. http://www.mbayaq.org (accessed September 12, 2007).

National Center for Atmospheric Research. http://www.ncar.ucar.edu (accessed September 12, 2007).

National Geographic Magazine. http://www.nationalgeographic.com (accessed September 14, 2007).

National Oceanic and Atmospheric Administration. http://www.noaa.gov(accessed September 1, 2007).

National Park Service. http://www.nps.gov (accessed September 14, 2007).

National Park Service, Katmai National Park and Preserve. http://www.nps.gov/katm/ (accessed August 9, 2007).

National Science Foundation. http://www.nsf.gov (accessed August 9, 2007).

Nature Conservancy. http://www.nature.org (accessed September 5, 2007).

The Ocean Environment. http://www.oceansatlas.com/unatlas/about/physicaland chemicalproperties/background/oceanenvironment.html (accessed September 1, 2007).

Oceana. http://www.americanoceans.org (accessed August 14, 2007).

"On Peatlands and Peat." International Peat Society. http://www.peatsociety.fi (accessed July 13, 2007).

Ouje-Bougoumou Cree Nation. http://www.ouje.ca (accessed August 25, 2007).

Scientific American Magazine. http://www.sciam.com (accessed September 5, 2007).

"Status and Trends of Wetlands of the Counterminous United States from 1998 to 2004:" U.S. Fish & Wildlife Service. http://wetlandsfws.er.usgs.gov/status_trends/National_Reports/trends_2005_report.pdf (accessed July 13, 2007).

Swedish Environmental Protection Agency. http://www.internat.naturvardsverket.se (accessed August 25, 2007).

Thurston High School: Biomes. http://ths.sps.lane.edu/biomes/index1.html (accessed August 9, 2007).

Time Magazine. http://time.com (accessed August 14, 2007).

"The Tundra Biome." University of California, Museum of Palentology. http://www.ucmpberkeley.edu/exhibits/biomes/tundra.php (accessed August 9, 2007).

U.N. Atlas of the Oceans. http://www.cmc-ocean.org (accessed September 12, 2007).

UNESCO. http://www.unesco.org/ (accessed September 5, 2007).

University of California at Berkeley. http://www.ucmp.berkeley.edu/glossary/gloss5/biome/index.html (accessed July 13, 2007).

U.S. Fish and Wildlife Services. http://www.fws.gov/

USDA Forest Service. http://www.fs.fed.us (accessed August 22, 2007).

"Wind Cave National Park." The National Park Service. http://www.nps.gov/wica/ (accessed September 14, 2007).

Woods Hole Oceanographic Institution. http://www.whoi.edu (accessed May 14, 2008).

The World Conservation Union. http://www.iucn.org (accessed August 22, 2007).

World Wildlife Fund. http://www.wwf.org (accessed August 22, 2007)

World Meteorological Organization. http://www.wmo.ch (accessed August 17, 2007).

Index

Numerals in italic type indicate volume number. Bold type indicates a main entry. Graphic elements (photographs, tables, illustrations) are denoted by (ill.).

Fragmentation reproduction, *2:* 269, 291–92;
  3: 436
France
  Normandy beach invasion of WWII, *3:* 392
  truffles, *1:* 94
Franklin, Benjamin, *2:* 264
Freezing
  lakes and ponds, *2:* 234
  mosquitoes and, *3:* 439
  tundra soil, *3:* 430, 432
  See also Ice
Freshwater. *See* Rivers and streams; Wetlands
Freshwater lakes and ponds, defined, *2:* 216
Freshwater marshes, *3:* 471–73
Freshwater swamps, *3:* 471
Friction, river/stream velocity and, *3:* 343
Fringing reefs, *1:* 59
Frogfish, *2:* 259 (ill.)
Frogs
  coniferous forests, *1:* 25
  deciduous forests, *1:* 102
  deserts, *1:* 142–43
  grasslands, *2:* 189, 190, 203
  lakes and ponds, *2:* 232, 232 (ill.)
  rain forests, *2:* 315–16, 316 (ill.)
  rivers and streams, *3:* 358
  wetlands, *3:* 484
Fulani people, *1:* 156; *2:* 215
Fungi
  coniferous forests, *1:* 15
  deciduous forests, *1:* 94–95, 94 (ill.),
    99–100
  deserts, *1:* 134
  grasslands, *2:* 183
  lakes and ponds, *2:* 227
  rain forests, *2:* 307–8
  rivers and streams, *3:* 353
  seashores, *3:* 400
  tundra regions, *3:* 434
  wetlands, *3:* 478–79
Fur
  chinchillas, *3:* 447
  musk oxen, *3:* 447
  tundra mammals, *3:* 444

G

Gagnan, Emile, *2:* 288
Galls, tree, *2:* 188
Ganges River
  ancient civilizations, *3:* 366
  delta, *3:* 350
  elevation, *3:* 351
  overview, *3:* 377–78
Gases, atmospheric. *See* Atmosphere
Gaspé Peninsula, *3:* 460–61
Geese, Greenland white-fronted, *3:* 491
"General Sherman" (tree), *1:* 8
Geography
  coniferous forests, *1:* 11–14
  continental margins, *1:* 54–60
  deciduous forests, *1:* 91–93
  deserts, *1:* 126 (ill.), 128–33, 128 (ill.),
    131 (ill.)
  grasslands, *2:* 180–83
  lakes and ponds, *2:* 222–25
  ocean floor, *2:* 265–68
  rain forests, *2:* 305–6
  rivers and streams, *3:* 346–51, 347 (ill.), 349
    (ill.), 350 (ill.), 351 (ill.)
  seashores, *3:* 394–99, 398 (ill.)
  tundra regions, *3:* 430–33, 431 (ill.), 432 (ill.)
  wetlands, *3:* 476–77
Geothermal energy, *2:* 218
Germany
  Normandy beach invasion of WWII, *3:* 392
  peat loss, *3:* 494
Geysers, *2:* 218
Giant anteaters, *2:* 202
Giant armadillos, *2:* 202
Giant otters, *3:* 374
Giant redwood trees, *1:* 7–8
Giant saguaro cacti, *1:* 136
Giant sequoia trees, *1:* 7–8, 11
Giant tubeworms, *2:* 274
Giant water bugs, *2:* 231
Gila monster lizards, *1:* 144, 145
Ginkgo trees, *1:* 111
Girdling, trees, *1:* 35

M